•••

PRESENCE IN THE FLESH

PRESENCE IN THE FLESH

•••

THE BODY IN MEDICINE

KATHARINE YOUNG

HARVARD UNIVERSITY PRESS
CAMBRIDGE, MASSACHUSETTS
LONDON, ENGLAND
1997

Printed in the United States of America

Library of Congress Cataloging-in-Publication Data

Young, Katharine Galloway.
Presence in the flesh : the body in medicine / Katharine Young.
p. cm.
Includes bibliographical references and index.
ISBN 0-674-70181-X (alk. paper)
1. Body, Human—Social aspects. 2. Body, Human—Symbolic aspects.
3. Medicine—Practice. 4. Medicine—Philosophy. I. Title.
[DNLM: 1. Human Body. 2. Symbolism. 3. Social Environment. GN 298 Y73p 1997]
GN298.Y68 1997
306.4—dc21
DNLM/DLC
for Library of Congress 96-39721

To Leonard J. Perloff, M.D.,

my companion and fellow adventurer in medicine

CONTENTS

• • •

ACKNOWLEDGMENTS

•••

When I presented my paper "Deciphering Signs of Presence in the Flesh: Surgical Incisions" at the American Folklore Society meetings in Boston, Shelley Posen told me I had a title for my book: "Presence in the Flesh." What follows took shape from that phrase.

The Wenner-Gren Foundation for Anthropological Research provided a grant that made it possible to conduct some of the fieldwork on which the study is based.

Daniel Lefkowitz produced useful transcriptions of audiotapes of gynecological examinations and surgical operations, despite his unfamiliarity with their exotic terminology and paraphernalia. The transcriptions in Chapters 2 and 3 are reworkings of his originals. Leonard J. Perloff, M.D., provided thoughtful clarifications.

Portions of several chapters first appeared in other forms: Chapter 1 draws on two papers, "Disembodiment: The Phenomenology of the Body in Medical Examinations," which appeared in *Semiotica* 73:1/2 (1989), published by Mouton de Gruyter, a division of Walter de Gruyter & Co., and "Narrative Embodiments: Enclaves of the Self in the Realm of Medicine," which appeared in *Texts of Identity,* edited by John Shotter and Kenneth Gergen (London: Sage Publications Ltd., 1989); Chapter 4 is based on "Still Life with Corpse: The Management of the Grotesque Body in Medicine" in *Bodylore* (published by the University of Tennessee Press, 1993), which I edited; and the coda appeared in an initial draft as "Perspectives on Embodiment: The Uses of Narrativity in Ethnographic Writing" in *Journal of Narrative and Life History* 1:2/3 (1991). I am grateful to the publishers for granting permission to make use of revised versions of these. The epigraph from Roland Barthes' Journal is excerpted from

Roland Barthes, translated by Richard Howard, translation copyright © 1977 by Farrar, Straus & Giroux, Inc.; reprinted by permission of Hill and Wang, a division of Farrar, Straus & Giroux, Inc.; also reprinted by permission of Macmilllan, Ltd. The epigraph by Gabriel Marcel is from his book *The Mystery of Being,* vol. 1; copyright © 1960 by Regnery Publishing, Inc.; all rights reserved; reprinted by special permission of Regnery Publishing, Inc., Washington, D.C.

The manuscript underwent a sea change in the wake of acute criticisms by two readers for Harvard University Press. I would also like to thank Elizabeth Gretz, the familiar spirit who coaxed and chivied it into materializing.

I would like to thank the physicians and patients at the pseudonymous "University Hospital" who good-heartedly participated in this study. By the rules of the game, they remain anonymous. I also appreciate the clear-mindedness of colleagues who make inquiry adventurous, especially Ken Gergen, Barbara Kirshenblatt-Gimblett, Elliot Mishler, and Amy Shuman. And I acknowledge the high-spirited interchanges with my students in philosophy of medicine at the University of Delaware that made teaching the body in medicine a pleasure.

PREFACE

• • •

PRESENCE, FLESHLY AND OTHERWISE

W*here am I?* There, behind the lineaments of those philosophers who start things off in Chapter 1. You hear me speak with and against them. As the hospital materializes over the course of these pages, I condense into its architecture as the perceiving eye. I traverse the space of inquiry. The patient turns up. The doctor. As I trail around after them, I am in the scene you see. Or in the wings, or with the audience, or directing, or editing, or writing. Every now and then, you may catch a glimpse through the text of a prior traversal, deforming this one, informing it, prefiguring and fading out of it: my earlier trudge, weighted with tape-recorder and notebook, guised in doctorate and dress clothes, riffling consent forms. How did that quest start?

The quest started with stories, my husband the surgeon's stories of adventures in medicine, medical mysteries, heroic ventures, legendary characters, all pervaded by the mythos of medicine, all narrative genres that transported me to another realm. These disembodied jaunts were succeeded by my own hunt for points of entry, searches through phenomenology, the anthropology of the body, bodylore, for keys to the realm of medicine; followed by negotiations with the university committee for research on human subjects; then with the chairs of departments of medicine, with physicians, and with patients for permission to penetrate the mysteries.

And all these are themselves preceded by what? By a history of thought that is and is not a lore of the body. I have come to this posture of attentiveness, of reserve, of implication or skepticism, out of the philosophy of mind in which I was first immersed academically, an immersion, not quite bodily, that nonetheless perfuses my tissues. And of course I succumbed to a romance of the Other. In this, physicians and patients allude to their forebears in anthropology, the Other, the exotic, the primitive. Medicine is at once a condensation of the ordinary and an alien domain. I see its practices, its jargon, its costumes and accoutrements both as cultural commonplaces and as tribal customs. Physicians appear mundane and exotic. If I set myself apart from them and fabricate difference, they become the Other, objects of desire, tantalizing me. If I enter into our kinship, I find us, the physicians, the patients, and me, jointly fabricating a world. And I have brought to the inquiry a bodylore of my own invention. And here philosophy issues from the body. Those prior selves, those embodiments and disembodiments, and others, earlier, elsewhere, shape this text.

Sometimes I announce my bodily presence in the scene as ethnographer, when I speak to the professor about Auschwitz in Chapter 1, for instance. Sometimes I become palpable as perceiver, as when I filter the patient's experience through my own sensibilities in the Coda. How could it be otherwise? Who else's body could I perceive through? But there is another body, the Other's body. I climb into it at the end of the Coda and embark on further investigations otherwise embodied.

And elsewhere, especially in Chapter 2, I vanish, yet bodily sensibilities, perceptual modalities, materialize multiple realities. My body, your body, splits apart.

At other times in the Coda, I mention myself as writer, and then you see not the adventurer of the body, but the adventurer of the mind. And ongoingly, over the course of the book, I am transmuted into the phenomenological "I," not my idiosyncratic self but the locus in the text for embodied perception. I position myself as the receptacle, the opening, the hold in the text for your body. Do come, hunker down, have a look around. We have to put on gowns and gloves, masks and foot covers, for surgery in Chapter 3.

And the patient, that obdurate massy body, has it receded from us? Or can I withdraw with it, relinquishing my bodily holds, into sleep, into death? Come on, you, too. We are with the corpse in Chapter 4. Why am I aware with a visceral twinge of the abdominal incision? It is the mystery of incarnation, the puzzle with which we have been grappling for some two thousand years, of the presence of the spirit in the flesh. That god, that God incarnate, is a representation of the mystery of ourselves. I shall speak to Descartes about it in the Conclusion.

But I have slipped in before Chapter 1, in the Introduction, where I exhort you to consider how nakedly the body goes into medicine. And before that, in the Acknowledgments, where I sketch the skeleton of a corpus of Others, known to you or not. And even before that, in the ghost of somebody, one other body, materialized in the dedication. And of course here, in the Preface, where I am meant somehow to come forward in my own person, though not in the flesh, to say where I come into this writing, this thinking, this inspecting. That is the question all prefaces address. And fail to answer.

I cannot put forward a history of the body as though it naturally grounded the production of the text. Rather, the body unnaturally invades the text, pouncing out here, poking through there, lurking behind this, or just beyond that. But you can find the body. Because this is its trace. And the body you find will be mine and not mine. It will be myself as you reconstitute me out of the imprints of my passage through the text, as you would reconstitute me anywhere, out of my corporeal and incorporeal clues. *Where am I?*

A presence can, in the last analysis, only be invoked or evoked, the evocation being fundamentally and essentially magical.

—Gabriel Marcel, *The Mystery of Being*

INTRODUCTION

• • •

THE PHENOMENOLOGY OF THE BODY IN MEDICINE

Medicine inscribes the body into a discourse of objectivity. The body is materialized even as the self is banished, creating that disjunction which is the core of medical phenomenology: the mind/body problem. In the realm of the ordinary, the body is the self, the site of my experiences, the fulcrum of my movements, the source of my perspectives. I experience myself as embodied. In the realm of medicine, the body is rendered an object. It is inspected, palpated, poked into, cut open. From being a locus of self, the body is transformed into an object of scrutiny.

This scrutiny is quintessentially performed in medical examinations, in the course of which the oddity of the body's ontological status becomes apparent. Consider what transpires. I walk into a room, take off all my clothes, and permit a perfect stranger to inspect my naked body. This despite considerable psychological and occasionally physical discomfort. How does this come about? Why isn't it a trespass, a violation, an assault, at least an impropriety? It goes against our customs, sets aside the practices of civilized life, flouts propriety. What transformation, what reconstitution, of my sense of my self and my body allows this unlikely intimacy, the physical examination?

Medicine does not articulate an etiquette of the body. Rather it unreflectively inhabits the sort of belief system folklore is uniquely designed to ferret out. The inquiry's root in bodylore, the study of the symbolic practices that make the body what it is supposed to be, is apparent in its attention to the body's constructedness. The body is not given as the physiological substrate on which medical discourse is mounted; rather, it is invented and transformed by medical discourse. Medicine, it becomes apparent, fabricates a body. It enacts its own mythos in routine practices that are properly investigated under the aegis of folklore, the study of forms of thought that have become typified within a discourse. Anthropologically, medicine is a culture with its own language, gestures, customs, rituals, spaces, costumes, and practices. Within medical culture, the body becomes the locus that corporealizes culture, enculturates bodiliness. The epistemology of the body in orthodox medicine, like the epistemology of the body in alternative medicine or in other cultures, displays a phenomenology, a panoply of modes of being specific to medicine. These give rise to a semiology of the body, decipherable in its proxemics,[1] kinesics, and sociolinguistics. Medical texts, medical talk, and medical tales, retrieved through discourse analysis, conversation analysis, and narrative analysis, adumbrate the body in medicine.

To inspect the medical body, I submitted a proposal to the Committee on Studies Involving Human Beings at a university on the eastern seaboard. The committee ensures the propriety of research ranging from drug trials at the university hospital to anthropological research in Micronesia. Mine fell somewhere in between. From some ethereal region, the committee issued me a dispensation. One of its requisites is that I am obliged to keep the names of patients, physicians, and the hospital itself confidential. Hence I have used the pseudonym "University Hospital." The names I have given the patients and physicians are fictitious, too, but they retain a tinge of the ethnicity of their originals where that was apparent.

Between 1984 and 1987, I was apprenticed in turn to two internists, one gynecologist, one surgeon, and one pathologist. I spent shorter spells with a second gynecologist, three other surgeons, two anesthesiologists, two other pathologists, several nurses, and a technician. My habit was to grab a notebook, hitch a tape recorder to my belt, and follow the physician through interactions with whichever patients had agreed to participate in the study. Over the course of that time, I observed and tape-recorded some sixty-nine internal examinations, thirty-nine gynecological examinations, nine surgical examinations, nine surgical operations, and three pathological examinations.

The resulting forty-one tapes of the medical examinations, the gynecological examinations, and the surgical examinations came out very clearly. The clarity was fortuitous, because I consider the examination the quintessential gesture of medicine. The minimalism of the pathology tapes and the multilayered complexity of the surgery tapes, however, both presented problems of retrieval. Pathological examinations were nearly impossible to hear for the curious reason that everyone in the morgue whispered, and nothing ever turned up on tape. My solution was to describe a choreography of movements, illuminated by occasional remarks. Surgery, in contrast, was so noisy that everything turned up on tape but nothing was intelligible. After the first few misadventures, I was able to foreground certain aspects of the conversation by clipping a tiny microphone to an amiable surgeon's collar. The result was that I attended to talk around the chief surgeon as the pivotal figure, one vein of a ramifying discourse enfolded in an architecture of gestures, moves, and spaces. Transcription conventions are appended in the Note on Transcription.

Getting consent for my activities was also tricky. The most delicate occasions to arrange to see were physical examinations, which patients perceived as personal. As I got more accustomed to talking to patients, they got more comfortable about participating in the research. After a while I was able to guess with some accuracy who was going to consent and who was going to refuse. But I was still often surprised. Characteristically, slightly over half the patients I asked on a given day consented, but sometimes, unaccountably, almost everybody consented and on one notable afternoon, nobody did. Did I deduce these differences or did I induce them?

My effect on these occasions was unpredictable. After the first few examinations, or the first half-hour of a longer procedure, the physicians seemed to get used to me. Because this was a teaching hospital, they were accustomed to being followed about by apprentices, or even eminences, of various sorts. But for the patients, I was often a new wrinkle. They may, for instance, have behaved more properly in my presence. But this propriety can only have enhanced the etiquette of touch I was after. Some of the patients were used to this sort of thing, and when I started to explain my research, they would say, "Oh, another one of those," and either wave me off or grab the consent form and sign it unceremoniously.

In both surgical procedures and pathological examinations, the patients had, as it were, absented themselves from the occasion. Getting consent from patients to see their surgery was often easy: patients do not regard surgery as personal. The difficulty was that I had to get additional consent not only from the surgeon but also from the rest of the operating team, which might consist of a resident, an intern, a scrub nurse, a circulating nurse, and an anesthesiologist. In the case of transplants, this was doubly difficult as the team was assembled when the hospital got an organ, often at three in the morning. In pathological examinations, the patient was obviously so profoundly absent as to be disempowered to consent. In that case, consent issued from the chief pathologist, who undertook to protect the corpse from perverse curiosity.

In the course of the research, the Wenner-Gren Foundation provided me with a grant with which, among other things, I hired a linguist, Daniel Lefkowitz, who transcribed a number of the gynecology and surgery tapes. He worked on the order of thirty hours to produce the transcription of a single surgical procedure. The 293 pages of transcription he made, along with my own indices and transcriptions of the twenty-six tapes of medical examinations, vastly facilitated my analysis of the ontological status of the body.

Instances of embodiment and disembodiment are inspected in Chapter 1, which examines how the body is transformed from a social subject into a medical object during an examination. In effect, I undergo a species of disembodiment. This creates the estrangement between self and object that makes the body problematic in medicine. Sequestering the self is intended to protect the social person from the trespasses of the examination, but it can have the effect of dehumanizing the body. It is possible, in response, to create an alternative locus for the disembodied self: the realm of narrative. Patients whose bodies are given over to the examination can carve an enclave out of the realm of medicine in which to reconstitute a self narratively. Narrating a self preserves the conventions of medicine with respect to disembodiment and at the same time provides a footing for the self on the occasion of the examination. The interplay between the medical body and the narrative self loosens the presumptions of both bodiliness and selfhood, giving examinations in which it transpires a postmodern cast.

Over the course of medical procedures, the body shifts between objective and subjective discourses. During a gynecological examination, both discourses are inscribed on the body at once. A discourse of social subjectivity, mapped onto the upper outer body, engages it visually; a discourse of medical objectivity, mounted on the lower inner body, engages it tactually. Olfactory and auditory perceptions transgress the boundary between these realms of discourse. Chapter 2 investigates how perceptual modalities make and unmake a world. Laminating modalities together inveigles the patient into substantiating two realities corporeally. She becomes a split subject whose pain can be hidden from one reality in the other. The corporeal and incorporeal footings attributed to or taken up by the body in medicine break the medical body out of its discourses and, at the same time, break the discursive body out of postmodernism.

The disposition to inscribe the body into a discourse of objectivity in the realm of medicine is set against the disposition to inscribe the body into a discourse of subjectivity in the realm of the ordinary. To regard the body as a locus of self is to conjure up a humanistic ideology out of the materiality of the body. One of the consequences of this materialization of a bodily self is that subjectivity comes to be represented in the humanist tradition as individuality. Thus surface inscriptions are taken to disclose the lineaments of individuality on the body. Even as the surface takes the marks of the individuated self, the body's interior is invoked in humanist thought as the site of an etherealized self. Chapter 3 considers the surgical incisions that overwrite the exterior and expose the interior of the body, calling into question both exteriorized and interiorized discourses of subjectivity.

The status of the body as object in medicine is figured in the corpse. Here the sheer materiality of the body might be supposed to exclude the self. But even eviscerated and dismembered, the dead body fails to reduce itself to its object status. Chapter 4 examines intimations of personhood on the dead body. The objectifying gaze of medicine finds itself deflected by the grotesque body of pathology. The history of medicine is the history of a mystification, a banishment, a disappearance of the grotesque body.

The properties with which medicine imbues the body are condensed versions of the properties of the post-Cartesian body to which we are all heir. An insubstantial self is split off from a solidified body. The body is conceived as an object; the self is conceived as a rarified effluvium haunting the object. As the Conclusion argues, the resultant disparity or incommensurability between the ethereal and material substances of the person make possible the move toward disembodiment in medicine. But the contrast in question is between neither mind and body, in which body is implicitly object, nor between subject and object, in which body is presumptively subject. The proper contrast is between body-as-self and body-as-object, and both of these are aspects of the experience of being a body. I am aware of my body, by turns, as self and as object. Medicine takes up or puts forward my body-as-object without necessarily, always, or altogether, and sometimes not at all, dispossessing me of my embodied self.

The bearing of narrativity on constructions of the body becomes pivotal in the Coda. "To write the body," as ethnographers, is likewise a narrative gesture. Conventions of perspective and voice fabricate the bodies of others and the universes of discourse they inhabit. These conventions are anchored in the bodies of ethnographers as percipients. Hence our cultural hierarchy of modalities of perception informs our social scientific epistemology. Realistic writing constructs the bodies of others as objects, separates their universes of discourse from ours and so estranges the ethnographer from the Other. This discursive move toward objectivity reiterates medicine's inclination to regard its subjects as objects. Unconventional writings, by contrast, fret the continence of the body, the closure of the universe, and the detachment of the perceiver. They reinscribe the body into discourses of subjectivity. The Coda invents different narratives in order to conjure up different embodiments.

These investigations of the phenomenology of the body in medicine proceed from the conscious body to the split consciousness of the gendered body through the unconscious body to the dead body. In each of these transformations, bodylore brings out the post-Cartesian metaphysics by virtue of which medicine produces the body. The material but insensible medical body is excoriated to expose its metaphysical sinews, attached, just beyond the site of incision, not to the bone but to the imagination. Medicine issues from a folk epistemology of the body. Despite the transformations of the body by medicine, signs of presence press through the flesh.

ONE

• • •

DISEMBODIMENT • INTERNAL MEDICINE

The immediacy of my experience of corporeality should be understood as an indication of the interior perspective I occupy with respect to 'my body'. I am neither 'in' my body nor 'attached to' it; it does not belong to me nor go along with me. *I am my body.*

—*Maurice Natanson, The Journeying Self*

I am inserted into the world bodily and my experience of the world comes to me through my body. The phenomenologist Maurice Natanson writes, "my body is the unique instrument through which I experience my insertion in the world" (1970, 17). My body is the locus of my percipience, the vantage point from which I perceive the situation in which I find myself. "The perspective from which and through which that situation presents itself is the insertion of the individual at some place in the social fabric" (Natanson 1970, 60). Not only do I apprehend the world corporeally but my body is also an aspect of the world, an object in it. "Visible and mobile," writes Maurice Merleau-Ponty, "my body is a thing among things; it is caught in the fabric of the world, and its cohesion is that of a thing. But because it moves itself and sees, it holds things in a circle around itself. Things are an annex or prolongation of itself; they are incrusted into its flesh, they are part of its full definition; the world is made of the same stuff as the body" (1964, 163). Here is the source of the temptation to take my body's objectivity as its paradigmatic condition and to suppose its subjectivity to be secreted invisibly inside. This might be called the physical object hypothesis of persons. "All the signs of mundane reality lend implicit support to the assumption that the model of the physical object in the quantified space of nature is a paradigm for the being of man in the world" (Natanson 1970, 2). The difficulty with the physical object hypothesis is not the materiality of the body but the resistance of its materiality to a putatively immaterial self. This dispossesses me from my corporeal self, as if I were different from my body.

If I am not dispossessed from my body, neither am I subsumed by it. The way I experience my body, the way I speak of it and think about it, is rooted not so much in its sensible apprehensions as in my symbolic perceptions. I learn my body even as I learn bodily. The anthropologists Shelden Isenberg and Dennis Owen write:

> The individual's body is presented to him, taught to him by society, usually in the manifestations of parents, and then by peers, perhaps also by schools. Our attitudes about our bodies arise from society's image of itself. So if we can learn how a person understands the working of that complex system called the body, its organization, its spatial arrangement, and its priorities of needs, then we can guess much about the total pattern of self-understanding of the society, such as its perception of its own workings, its organization, its power structure, and its cosmology. The human body, then, is a universal symbol system: every society attempts in some way to socialize its members, to educate its bodies. (1977, 3)

This is not a matter of imposing education on the resistant flesh. A natural fact, a physiological given, a material substrate is not somehow prior to the idea of the body. That notion is grounded in understandings which, Barbara Duden argues, "implicitly presuppose something like a nonhistorical (biological) matter of the body, which is then molded by time and class, on which 'culture is imprinted' or which is 'culturally shaped.' The matter itself always remains a given" (1991, 6). The naturalization of the body is a consequence of a history, Duden points out, in the course of which "the body and its environment have been consigned to opposing realms: on one side are the body, nature, and biology, stable and unchanging phenomena; on the other side are the social environment and history, realms of life subject to constant change. With the drawing of this boundary the body was expelled from history, and the problem of how it has been perceived fell outside the sphere of social history" (1991, vi). But the body bears its historicity materially. It is only perceptible as it is conceived. My body, as it were, inhabits its own subjectivity. It is imprinted from the start with traces of my being in the world, of my language, my culture, my experience, how my body is handled and the interpretation I put on that handling, so that, as Merleau-Ponty writes, "whenever I try to understand myself the whole fabric of the perceptible world comes too, and with it come the others who are caught in it" (1964, 15). Natanson expands this: "The world I inhabit is from the outset an intersubjective one. The language I possess was taught to me by others: the manners I have I did not invent; whatever abilities, techniques, or talents I can claim were nourished by a social inheritance; even my dreams are rooted in a world I never created and can never completely possess" (1962, 103).

The intelligibility of the world is constituted corporeally. The body is the source as well as the site of symbolic understandings. Maxine Sheets-Johnstone writes:

> Meanings are not free-floating entities; meanings are incarnated, anchored in living bodies. It is clear why corporeal representation is a fundamental biological matrix. It is a primary mode of symbolization and communication. Where meanings are *represented,* animate bodies represent them corporeally. In their form and behavior animate bodies are potential semantic templates. This is why a psychology, aesthetics, archaeology, and linguistics of symbolizing behaviors is possible—why pears and mountains can represent female breasts and umbrellas and tree trunks can represent penes. (1990, 121)

The body symbols that inform culture are already replete with mentality as well as corporeality. Victor Turner describes this:

> The cosmos may in some cases be regarded as a vast human body; in other belief systems, visible parts of the body may be taken to portray invisible faculties such as reason, passion, wisdom and so on; in others again, the different parts of the social order are arrayed in terms of a human anatomical paradigm. (1970, 107)

By the same token, Mary Douglas argues, incorporeal as well as corporeal properties are already inherent in the symbolic body.

> The physical body is a microcosm of society . . . At the same time, the physical body, by the purity rule, is polarized conceptually against the

> social body. Its requirements are not only subordinated, they are contrasted with social requirements. The distance between the two bodies is the range of pressure and classification in the society. A complex social system devises for itself ways of behaving that suggest that human intercourse is disembodied compared with that of animal creation. It uses different degrees of disembodiment to express the social hierarchy. The more refinement, the less smacking of the lips when eating, the less mastication, the less the sound of breathing and walking, the more carefully modulated the laughter, the more controlled the signs of anger, the clearer comes the priestly-aristocratic image. (1973, 101)

Hence, as this interplay between philosophers and anthropologists suggests, a phenomenology of the body also recovers the folklore that is invested in it. Properties can be seen in body symbols or in the symbolic body, in the way cultural inscriptions become visible in the body and in the way corporeal inscriptions become visible in culture. Such inscriptions body forth properties held to be present in the flesh.

> The social body constrains the way the physical body is perceived. The physical experience of the body, always modified by the social categories through which it is known, sustains a particular view of society. There is a continual exchange of meaning between the two kinds of bodily experience so that each reinforces the categories of the other. As a result of this interaction, the body itself is a highly restricted medium of expression. (Douglas 1973, 93)

This inquiry opens with an investigation of the symbolic properties of the body in a situation in which bodily intimacies are routinely undertaken: medical examinations.

FRAMES AND BOUNDARIES

Because the body is invested with symbolic properties, its parts are treated differentially. "Indeed," writes Erving Goffman, "this differential concern tells us in part how the body will be divided up into segments conceptually" (1979, 38). What Goffman calls "evidential boundaries" are interposed between others and their visual, auditory, tactile, and olfactory apprehension of the body or its parts (1974, 215; see also 1959, 151–165, on communication barriers or boundaries; 1976, 127, 140, on participation shields). The body itself constitutes an evidential boundary. States that are supposed to be internal to the individual "make their appearance through intended and unintended bodily expression, especially through his face and words. His epidermis can thus be seen as a screen, allowing some evidence of inner state to pass through, but also some concealment . . . In addition to functioning as a screen to what is presumably inside him, his body also functions as a barrier which prevents those on one side of him from seeing what is directly on the other side or those in back of him from seeing his facial expression" (Goffman 1974, 216).

Concealments of the body itself behind evidential boundaries are at issue here. These boundaries can take the form of bodily clothing; scents and de-scenters; voice levels and direction; gaze direction, concentration, and focus; position, posture, gesture, and movement; arrangements of furniture, spacing, and architecture. Such boundaries at once locate and conceal those parts of the body that are symbolically charged. Over the course of a medical examination, certain of these boundaries are peeled away to permit a close inspection of parts of the body. At the same time, other boundaries are introduced. Thus persons to be examined are put into a closed room so that its walls substitute as evidential boundaries for the clothes they take off, the difference being that the physician is inside the boundaries along with the person.[1] A complex choreography involving the disposition, shift, removal, and replacement of boundaries is undertaken by physicians in concert with their patients. The management of evidential boundaries during medical examinations is one of the concerns of this chapter.

For the purposes of the examination, the body is reframed to exclude some if its symbolic properties, especially sexual ones. Symbols, then, are not inherent in the body or its parts; rather, they are interpretations attributed to it by persons in situations.[2] Framing is accomplished by greetings, forms of address, language about the body, deference and dominance behavior, costuming, role play, the management of verbal and nonverbal delicacy, ritual, and metacommunication. These frames create and sustain alternate realms of experience, to adopt Alfred Schutz's terms (1973, 252–253), the realm of the ordinary in which I am a social person and the realm of medicine in which I am a body, and orchestrates the passage between realms. From being a locus of self, patients' bodies are transformed into objects of scrutiny, organs in a sack of flesh. Physicians' bodies, their personhood narrowed though not expunged, become instruments of detection, lodgments of a perceptual apparatus. The management of frames during medical examinations is, therefore, the other concern of this chapter.

Following John Locke ([1690]1959), a medical examination could be taken to operate under a social contract which would read: "for medical purposes, I grant license to this physician to examine my body." Under this interpretation, reframing the body as an object could be understood as the act John Searle describes as a performative (1969), a linguistic or metalinguistic message that enacts what it expresses and takes the reading, "I hereby render this/my body an object." The contract would be supposed to hold across the interaction, performing a transformation on the body that renders it, for the nonce, an object instead of a self. In practice, however, it becomes apparent that realm-shift is not a prior contract but an intermittent, periodic, or partial phase, layer, or aspect of the medical examination. The incompleteness of the transformation can be seen from the one side as the uncontainability or flooding of the patient's social person through the reframing, and from the other as the physician's invocation of or attestation to the presence of a self in the body. An appreciation of the contractual expectations of medical examinations does not in itself transform persons into patients; it merely makes them alert to the cues for their own transformation. A good deal of fiddling with the body's ontological constitution is required before the shift from one realm to the other is complete. Indeed, it may be that it is this ontological unsettledness that tinges the realm of medicine with its characteristic air of unease.

SPATIO-TEMPORAL ONTOLOGIES

Realm-shift in medicine is grounded in the distribution of realms in space and their sequencing in time. The University Hospital itself appears to seal the realm of medicine off from the outside world. But within the building, the differentiation between the realm of medicine and the realm of the ordinary is in some measure reinvoked. A network of public spaces and connecting pathways interlaces the network of professional spaces and connecting pathways, with points of intersection and areas of transition between systems. Crossing over from the realm of the ordinary into the realm of medicine entails a change of ontological condition. Physicians are regarded as initiates of the realm of medicine, and pass freely across the border. Other members of the realm of medicine can be required to display tokens of their status. Nurses pass across without question by virtue of their uniforms. Other employees of the hospital (cooks, cleaners, maintenance people, mail personnel) who are not native practitioners may wear badges that mark them as insiders. Patients, who are outsiders, must undergo a transformation in order to become participants in the realm of medicine. The routines associated with conducting medical examinations can be regarded from this perspective as rituals for effecting this transformation.[3]

The maintenance of physical boundaries between spaces supports the maintenance of conceptual boundaries between realms, for, as Douglas writes, "Any structure of ideas is vulnerable at it margins" (1970, 145). To maintain discretion between realms despite the trafficking across their borders, crossover points are narrowed, partially obstructed, concealed, or sealed. Where the integument between realms is thin, discretion may be maintained by locked doors to which staff people have keys. Where the pathway between realms is open, it may be narrowed to a corridor or partially obstructed by desks. At these points, guardians of the realm may be posted in the form of receptionists, secretaries, security guards, and the like, who monitor the ontological propriety of those passing through. This monitoring must be done with considerable caution. It is not proper, for instance, to challenge a physician—they are understood to be inextricably enfolded in their roles—so that anyone who makes the crossover with appropriate style is likely to be passed without question. For insiders, passage into the realm of medicine is eased by the existence of separate paths and entrances either obscurely placed or locked. For outsiders, passage between realms is slowed, obstructed, deflected, or sequentialized partly in order to provide interstices in which to accomplish transformations. Emergencies do not constitute breaches of the stringency of the system; for them, the tempo of the transfer is simply accelerated.

Within the realm of medicine, spaces are further differentiated. At University Hospital, practitioners of a given specialty are clustered together in separate suites in which the distinction between the realm of the ordinary and the realm of medicine is reinvoked. Internists, for instance, practice in the Department of General Medicine, consisting of a waiting room with reception desks set up beside the opening into the corridor that leads to physicians' examining rooms and offices, and further along to supply rooms, secretarial cubicles, the chairperson's office, and a conference room for medical staff. These deeper regions are never penetrated by persons in their role as patients. Persons as patients shift between the waiting room and the offices and examining rooms. Because of the nesting of networks within networks of spaces within the hospital, however, there is no single boundary between the realm of the ordinary and the realm of medicine. Instead, there is a modulation of ontological properties from outer to inner spaces. Commensurately, a person's movement through these spaces is sequentially ordered so that there is also no single moment of transition between realms. Persons are not turned into patients; rather, they undergo a series of transformations in the course of which they become patients.

REALM-SHIFT

Both physicians and persons initially carry their identities across the border into the other realm. Indeed, physicians retain the accoutrements of their medical role pervasively in the realm of the ordinary. It becomes a social presentation, attended by an analogue of the status it holds in the hierarchy of medicine. Physicians' sense of self is deeply invested in their professional role. Unlike individuals with more lightly held roles—professors, for instance—whose titles can be quite easily detached, physicians' roles are not lightly discarded, and their titles tend to remain attached in ordinary life.[4] This retention can be gracefully regarded or, conversely, uncomfortably perceived as a kind of aristocracy. By contrast, except in the case of the gravely or frequently ill, the patient role is quite transient.[5]

As persons become patients, they relinquish their social personae. They divest themselves of some of their social properties with their clothes. Taking off layers of clothing circumscribes the self by limiting its extensions into social space. The boundary of the self is not ordinarily coterminous with the skin. It extends not only to objects attached to the body but also to objects possessed by the person and to the envelope of space that contains them (Sommer 1969, viii; Hall 1969, 113–128, 177, 25–38; Goffman 1979, 28–60). Constraining the self to its bodily integument is a move toward rendering the body an object.

Occasionally, for certain kinds of examinations, patients are permitted partial retentions, in the form, for instance, of all their clothes (provided that their shirts can be rolled up at the sleeves or opened at the front), or of all their clothes from the waist down, or of some of their underwear, or perhaps just their socks. Conventional underwear, not being socially presentable, shifts from the realm of the ordinary to the realm of medicine fairly easily. Socks are quite another matter. It may be that feet never quite enter into the region of medicine unless it is they that are being examined. Interestingly, many patients whose feet are not being examined nevertheless take off their socks, apparently in the interest of completing the transformation from one realm status to the other. These remnants of their social appearances give patients a hold on themselves during the examination, but in so doing they create an anomaly within the realm of medicine by maintaining there some aspects of a presentation of another ontological status. Such partial transformations appear to be tolerable when the examination is not regarded as very extensive or intimate. Complete changes which appear to be functionally designed to allow complete access to the body are phenomenologically designed to allow the complete transformation of the body from self to object.

Physicians achieve a commensurate narrowing of self by the addition of a layer of clothing: a knee-length white coat.[6] The archeology of these artifacts is suggestive here: the layering of an outermost and predominant role over a complete social person as opposed to the reduction of a complete social person to a diminished role. In women who are patients, the circumscription of self is arrived at by both these means, the removal of layers of clothing and the addition of a white gown. Both physicians' and patients' garments reduce sexual signaling and other social communications by concealing the contours of the body. Thus at the beginning of the physical examination, women achieve a superficial parity with their physicians, but one that is easily undermined by the flimsiness of the gown and its lack of supporting undergarments. Men patients may be said to sustain a deeper parity with men physicians, but not one that is symmetrically displayed. The entanglement of professional with gender roles here is evidenced by women physicians who are exercised about the question of whether or not they should have men patients put on gowns for examinations. At issue is whether to interpret the examinations under the medical paradigm, in which the physician is regarded as gender neutral but the patients are supposed to display differential modesty, or under a gender paradigm, in which interactions within the same gender are regarded as neutral and interactions between different genders are regarded as sexually charged. Most women physicians in the hospital I studied provide women patients gowns and not men in an attempt to establish their gender neutrality.

Physicians make their initial appearance already in costume for their role, whereas patients change costume in the course of the performance. Persons who are physicians, then, present themselves in what is initially a social situation in the restricted role they expect to sustain throughout the interaction, thus constituting themselves what Alfred Schutz calls "enclaves" of another ontological status within the realm of the ordinary (Schutz 1973, 256n). Persons who become patients work a transformation on their own bodies from self to object and back to self again, thus shifting realms during the examination. As they move into the realm of medicine, the balance is reversed and persons become enclaves of the ordinary in the realm of medicine. Because enclaves of one realm thus extend into the other, the transformation is never complete.

The existence of multiple realm statuses in a single context creates ambiguity: alternative interpretations of a given event or object are always possible. This ambiguity can be used strategically to manage the examination, but it also gives rise to an uneasiness about how to move, especially by, and with respect to, the transmuting self (see Goffman 1974, 548, 560–575). For this reason, there is some impulse to achieve ontological congruity with the realm of medicine by turning the person into a patient. The rhythms of the examination seem to be designed to arrive at this unambiguous condition by the physical examination. Hence, at the center of the medical examination, the body of the patient is most nearly congruous with the realm of medicine, whereas at the periphery, it is most nearly congruous with the realm of the ordinary. The physician's presentation of self, in Goffman's phrase (1959), as a hold or transfix and the person's presentation of self as a variation or transform display the asymmetry between the modalities in which each of them moves through the examination. Multiple statuses are apparently easier to handle than changing statuses. Being in both realms provides a better grounding for interaction than being in neither. The interstitiality of transitional states eludes ontological placement (see Douglas 1970, 137–153; Leach 1971,

132–136; Van Gennep 1960, 1–14). It is to protect these interstices that changing clothes, which shifts the body from self to object, characteristically occurs behind evidential boundaries. The problem is not a sexual one but an ontological one.

The process of transformation might be said to begin on being born into a society with its particular conventions about the symbolic properties of the body. It is abstracted in the social contract, implicitly invoked in making the appointment, and put into operation on entering the hospital. Once there, persons feel obliged to observe some of the conventions of the realm of medicine: clothes are carefully chosen and arranged with an eye to the proprieties of the situation, cleanness perhaps being of more moment than formality, movements are contained and directed, voices are subdued. But it is in the waiting room that persons await realm-shift, and await, too, the cues that tell them when to shift realms. A person's social self is not held in abeyance, but put on the alert for these cues to its own transformation. This inquiry cuts in on the unfolding of such cues from the point in space or moment in time when the person first sees the physician face-to-face.

THE REALM STATUS OF THE BODY IN INTERNAL MEDICINE

In the Department of General Medicine, people come into the waiting room and give their names to one of three receptionists whose desks are arranged so that they form an extension of the corridor into the inner rooms and guard its entrance. Having given their names to an intermediary, persons then wait to be reinvoked as patients. In this practice, physicians collect the names of their next patient from the chart that is delivered to a rack outside their door by the nurse or from the receptionist or her list and come out into the waiting area and call the name out, scanning the waiting room for responses. They appear already accoutred for their role in white coats, but this appearance can be inflected with propriety or panache. Matthew Silverberg, M.D., wears his crisp coat over a well-cut three-piece brown tweed suit with a blue shirt and tie, and brown oxfords. The coat, which he wears open at the front, fits well across the shoulders and loosely over the body, effectively concealing its contours. It holds his stethoscope in the lower right front pocket and his pens in the upper left. He is a slender man with a narrow head, close-cropped hair, and glasses, and his face, appearing above the layers of coat, jacket, and shirt collars, has a withdrawn, fastidious air. Adam Kleinfeld, M.D., is a tall bearded man with glasses, a mild voice, and a slight stoop. His dark greyish hair is curly, longish, and undisciplined at the ends. He wears a grey jacket and trousers with a soft old sweater-vest and loafers. His stethoscope is hung around his neck, his coat is rumpled, and the short belt across the back of it has come unbuttoned. These modulations of the conventions of the realm of medicine provide each physician a slightly different foothold in the same realm.

Opening Frames

Greetings and farewells attest to the presence of social persons. Dr. Silverberg addresses his patients by a title of courtesy (Mr., Mrs., Miss) and a surname, and introduces himself by his surname and his professional title, which can have in this culture the aspect of a title of rank, thus inserting into the sociality of greetings a hierarchy of statuses. Dr. Kleinfeld addresses his patients by titles of courtesy and last names, by first names, or even by nicknames, but does not introduce himself. The hierarchical distinction between titled and untitled persons is to some extent preserved in the distinction between naming and what Goffman has called no-naming. In response to this informal naming system, his patients often call him "Doc." Each physician is thus solicitous of his patient's social person but dominant over it.

These moves for differentiating doctor from patient can be confounded by the presence of a patient who is also professionally titled. As he leans across his desk to shake hands with Michael Malinowski, Ph.D., a seventy-eight-year-old professor of Jewish history and literature, Dr. Silverberg uses the "sir" solution to show deference yet reserve title: "Hello, sir, how are you?" Shaking hands inflects touch as initially social and symmetrical, but requires Dr. Silverberg to transform the meaning of touch during the physical examinations to achieve the proper asymmetry and objectivity. Dr. Kleinfeld abstains from handshaking and so reserves touch for his professional attentions to the patient's body, but in so doing he keeps his social distance. What Dr. Silverberg creates is a realm of formality and social proximity, whereas what Dr. Kleinfeld creates is a realm of flexibility and social distance. By inserting dominance relations into social forms, both clusters of greeting behaviors operate to lodge control over the shift from the realm of the ordinary to the realm of medicine in the physician.

Spatio-Temporal Frames

Realm-shift is accomplished in part by moving from the waiting room to an office or examining room. Dr. Kleinfeld collects his next patient's chart from the rack outside his room, comes out into the waiting room and says, "Mrs. Peary?" On hearing her name, Mrs. Peary stands up, unfolding her body to view, and then moves into the physician's greeting space so that their apprehensions of each other are limited to the upper body and focused on the face. As she approaches him, Dr. Kleinfeld says, "Will you come in, please." He turns and precedes her down the hall to his room. He enters and crosses over to his desk, nodding at the chair beside it as he passes. She follows him in and sits down. With a patient he knows well, Dr. Kleinfeld says, "Hi, Dave," and waves him in. In contrast Dr. Silverberg comes out into the waiting room and says, "Mrs. Cenci, please." As she approaches, he leans toward her and says, "Dr. Silverberg is who I am," and shakes hands. She says, "Hi." He precedes her to his office, goes in ahead of her, pauses near a chair and says, "Have a chair, if you would," waits for her to sit, then says, "Thank you," goes around to the other side of his desk, and sits down. Or, with a patient he knows well, Dr. Silverberg says, "Mr. Rachelson, Hi." They shake hands in the waiting room and then Dr. Silverberg precedes him down the hall to his office.

These inner rooms are arranged at once to lay the patient open to the physicians's regard and to protect him or her from it. Dr. Kleinfeld sits in a chair facing his desk, which is flush against the wall. His patient sits in a chair set side-on to the desk, facing out into the room. Thus the desk does not interpose an evidential boundary between their bodies, and their bodily orientation and gaze direction are turned away from each other. Dr. Kleinfeld modulates this basic arrangement by flicking his swivel chair out and away from the desk, leaning back, and looking at his patient, or pulling the chair into his desk, leaning forward, and looking at the patient's chart. Dr. Silverberg, however, sits behind his desk, which faces out into the room. The patient's chair is drawn up sideways along the front so that the desk is interposed between the physician's lower body and his patient's, but he faces the patient across it and maintains eye contact. He modulates this basic arrangement by glancing down at the patient's chart on his desk, or turning aside to read a chart in his lap. In both instances, persons can turn their upper bodies to face the desk, thus achieving a side-to-face orientation with Dr. Kleinfeld or a face-to-face orientation with Dr. Silverberg. Four variables appear to work together here as a system: evidential boundaries, eye contact, personal spacing, and bodily orientation (Sommer 1969, 12–38; Hall 1969, 1–146; Goffman 1974, 215; Goffman 1979, 28–60). Concealing the lower body with an evidential boundary like the desk permits the maintenance of a face-to-face orientation with a high degree of eye contact at close proximity. Lowering evidential boundaries at about the same distance is associated with side-to-side orientation and reduced eye contact. Both constellations of arrangements maintain some evidential boundaries at close quarters.

With the shift of realms, persons become enclaves of the ordinary in the realm of medicine. The continued presence of a social person can be attested to by the physician, as when the professor, Dr. Malinowki, says to Dr. Silverberg:

Dr. M:	I will be in July the— seventy-nine.
Dr. S:	Seventy-nine
Dr. M:	July twenty-sixth I will be seventy-nine.
Dr. S:	You'd never know it. *((For transcription devices, see the Note on Transcription.))*

Attempts by the patient to attend to the physician as a social person can get short shrift. Dr. Kleinfeld escorts Mrs. Hardy to his examining room, goes over to his desk, and says:

Dr. K:	You got he— Come on you sit here— *((Motions to his desk. She sits. So does he.))* You got here while I was eating my half of a sandwich.
Mrs. H:	Oh just half a sandwich. *((Chuckles.))* You didn't eat a whole one huh.
Dr. K:	Now where were we. *((He turns to the patient's chart on his desk.))*

Though within the realm of medicine, these rooms are not ontologically pure. Photographs of his children, for instance, incline Dr. Silverberg's office toward humanity. Framed medical degrees would inflect it toward professionalism.

Conceptual Frames

Realm status is materialized in the architecture of these spaces because, as Gregory Bateson argues, "human beings operate more easily in a universe in which some of their psychological characteristics are externalized" (1972, 188). Realm-shift is, in fact, only incidentally spatial; it is essentially conceptual and can be modulated by the interactions of persons over time as well as by the movements of the body in space. The remark, "How are you?" can serve as a pivot between realms. In the realm of the ordinary, this turn of phrase occurs as a greeting formula to which the proper response is, "Fine." In the realm of medicine, it has the status of an inquiry about the patient's health, to which "fine" is not the proper response. Patients are instead supposed to respond with accounts of their medical condition. Dr. Kleinfeld says to Mrs. Frye:

Dr. K: So how are you.
Mrs. F: Same old thing mostly sick.

Since in the realm of medicine, the response to "How are you?" is not properly "Fine," producing this response can cause a slight hitch in the shift of realms. Dr. Silverberg says to Mrs. Johnson:

Dr. S: How are you.
Mrs. J: Fine.
((Mrs. Johnson then makes a joke about saying she is fine when she is not, noticing and thereby effecting the appropriate shift of realms.))

This hitch can also be handled by the physician, as when Dr. Kleinfeld says to Rose Shawn:

Dr. K: Hi.
How're you doing.

Mrs. S: I
[[[ ]
Dr. K: I guess I'm supposed to tell you.
Mrs. S: Right.

The catch with this form of inquiry is people's inclination to use a socially correct response instead of a medically informative one, resulting in a failure to shift realms.

To hedge against this failure, these physicians use three strategies. One is a transformation of the inquiry so as to make it less formulaic. Thus Dr. Kleinfeld tries, "How're you doing?" or "Hi, how've you been today?" The difficulty with this solution is that these phrases are easily taken as the greeting formulae they are intended to transform. A second strategy is to raise the specificity of the inquiry. When he turns to her chart, Dr. Kleinfeld says to Mrs. Hardy:

Dr. K: Now where were we.
Mrs. H: Well that was (not much)
((alluding to the sandwich)).
Dr. K: I thought you were having a cold.

A third strategy is to eliminate the ambiguous inquiry altogether and move directly into the medical realm. Dr. Silverberg says to Mrs. Cenci, after she sits down, "Now you're forty-three." The physician's orchestration of realm-shift can be confounded by persons who do not take on the role of patient. Dr. Silverberg had been examining Dr. Malinowski's wife, who is now resting on a litter in the waiting room. On taking his leave of her, Dr. Silverberg signals her husband, who is waiting there with their son, Alex, to come down the hall into his office. On the way, the nurse takes Dr. Malinowski aside to weigh him, and Dr. Silverberg goes into his office. When Dr. Malinowski comes in with Alex and the nurse, Dr. Silverberg goes round behind his desk where he stands and leans across to shake hands with his patient, saying "Hello, sir, how are you?" At the same time, his nurse says to him, "One ninety-nine," referring to the patient's weight, and goes out. Dr. Malinowski does not respond to Dr. Silverberg's proffered hand. Because it is clear to Dr. Silverberg that he and his nurse have spoken at the same time so that Dr. Malinowski might not have heard him, he attempts another greeting,

"Happy to meet you." At the same time, Alex says to his father, "Why don't you sit here," and gestures to the chair across the desk from the doctor. Dr. Malinowski still does not respond, so Dr. Silverberg, abandoning greetings, says, "Have a chair" and indicates the same one Alex has. The professor, looking around the office, says, "Where should I sit on?" Dr. Silverberg quickly puts in "Dr. Silverberg is who I am" at the same time that Alex is insisting, "Here—sit here." Dr. Malinowski, responding to neither of them, sees a chair behind the desk near Dr. Silverberg's and says, "Let me sit down here (so I can hear you)," at which point Dr. Silverberg, realizing Dr. Malinowski is quite hard of hearing, pulls the chair over next to his and says, "You want to sit here? All right." The professor gets settled saying, "(I sit) here," and Dr. Silverberg says, "Fine. Now this is your ( ) chart," turns to look at it, realizes it is the wrong chart and excuses himself to go out and get the right one.

The difficulties of this initial scene turn out to be grounded in the fact the patient is rather deaf. Parenthetical remarks and stage directions by the nurse and Alex are phrased as back-channel effects (Goffman 1981, 28), but occur simultaneously with the initial greeting sequence with Dr. Silverberg so that Dr. Malinowski cannot make out what the doctor says. For his part, the professor is casting about trying to work out an arrangement whereby he will be able to sit nearer the doctor in order to hear him. His failure to respond appropriately to either greetings or directions creates an initial misimpression that he is disoriented. When Dr. Silverberg returns with the correct chart, he proceeds to move into the realm of medicine with inquiries about Dr. Malinowski's health. Dr. Malinowski, sitting right next to him, responds crisply.

The Dislodgment of the Self from the Body

The realm of medicine is further differentiated into two lesser realms called by physicians the history-taking and the physical examination. The history-taking begins the dislodgment of the self from the body by turning the person's attention to her or his own body as an object. For instance, Dr. Silverberg verbally disarticulates the professor's body into parts, inquiring in turn about his height, weight, age, and health (whole body concerns); then about his eyes, throat, blood, heart, chest, finger, heart again, breath, ankles, and back (body parts, substances and processes); then about allergies, drinking and smoking, his relatives' diseases, marriages, and children (whole body concerns again); then his stomach, head, eyes, nose, throat, bowels, urine, stomach again, muscles, bones, and joints (parts, substances and processes). This verbal disarticulation loosens the person's investment of a self in the body insofar as the self is felt to inhere in the body as a whole, not in its parts. Shifting attention to the body as an object renders the person warden of her or his body.

The person's recognition of this move toward detachment is evident in the way Mrs. Hardy refers to her body during history-taking.

Dr. K: I thought you were having a cold.
Mrs. H: Yeah a cold and the I had this uh
back.
Dr. K: It seems
some difficulty with your back today=
Mrs. H: Yeah.
Dr. K: O.K.?
And where are you with all those things?
Mrs. H: Well
the cold got a little better
. . .
And then was there—
Oh the back
kept hurting . . .

Over the course of the history-taking, not only does Mrs. Hardy come to refer to her cold as "the cold," but she even refers to her back as "the back," as if she were somehow detached from them. This kind of objective self-referencing also informs her gestures. In describing pain in her back and belly, Mrs. Hardy arches in her chair, flares her hand out behind her back at waist level, then gestures over her belly, fingers still flared, not touching her body but directing attention to it with the propriety appropriate to an outside observer.

Occasionally it is difficult for the physician to induce a person to shift attention to the body as an object. Dr. Silverberg says to Dr. Malinowski:

Dr. S: How is your health.
Dr. M: I wouldn't complain.
Basically.

No realm-shift, so Dr. Silverberg tries a rerun.

Dr. S: If Alex hadn't asked you to come
in would you have?
Dr. M: I
had in mind I
needed a check-up.

Still no medical information, so Dr. Silverberg says,

Dr. S: Is there anything that's bothering you
other than the hearing.
Dr. M: Nothing.

Finally Dr. Silverberg formulates a medical inquiry out of an observation of his own. And he addresses it to the son rather than the father.

Dr. S: He's hoarse.
What's the history of that.

Alex tries to give some account of it, but they still fail to draw Dr. Malinowski into the realm of medicine, so Dr. Silverberg says,

Dr. S: Have you ever had any problem with your heart.
Dr. M: No.
Dr. S: No heart attacks?
Dr. M: Pardon me?
Dr. S: Heart attacks?
Dr. M: No.
Dr. S: No pain in the chest?
Dr. M: No pain in the chest.

Dr. Malinowski appears to have got the import of this line of inquiry, however, because as Dr. Silverberg starts to make another observation, Dr. Malinowski announces one of his own.

Dr. S: I
noticed that=
Dr. M: I am a graduate of Auschwitz

And he goes on to tell an anecdote that leads to an account of his tuberculosis, that is, a pain in the chest.

The Body as Object

If the history-taking is still a realm of the self, though one in which the self is becoming detached from the body, the physical examination is a realm of the body, and one in which the body is rendered an object. The physical examination is organized around acts directed to the body. Talk directed to the self is inserted into interstices between acts. During history-taking, by contrast, acts are inserted into interstices in talk. For instance, as he takes Dr. Malinowski's history, Dr. Silverberg asks him, "Do your ankles ever puff up?" When Dr. Malinowski appears uncertain, Dr. Silverberg says, "Tell you in a minute," reaches over to Dr. Malinowski, who is still sitting next to him, tugs down his sock with one hand and touches his ankle with a fingertip. Enclaves or strips from one realm are thus inset or interlayered in the other.

Realm-shift can be accomplished either by reframing the same space as a different realm, or by changing spaces. Dr. Kleinfeld combines his office and examining room in one space. His desk and chairs are clustered along one wall, the examination table, sink, and stools along the other. The spaces can be divided by a curtain running along the length of the room so that if a person is to change clothes, she or he can retire behind the curtain to do so. This flexible use of boundaries reframes the same space as a different realm. Dr. Silverberg keeps two separate rooms, an office for history-taking and an examining room for the physical examination. The ontological transformations of the medical examination are thus fitted into different realms already constituted by these fixed boundaries. The term "examination table" is indicative of the realm status of these regions. What might be seen from the person's perspective as a bed on which to lie is instead described from the physicians' perspective as a table on which to inspect things. As physician and patient move into intimate space for the examination, there is an enormous reduction of eye contact and face-to-face orientation along with the deployment of other evidential boundaries. The heightening of tactile and olfactory senses is accompanied by an elision of social presence.

The transition between realms is managed by reframing the body as an object. At the conclusion of the history-taking, Dr. Kleinfeld says to Mrs. Hardy,

Dr. K: O.K.
Right I want to
uh check your blood pressure but I want you to get undressed
and put the gown on so I can feel your belly
all right and I want to check you for blood—

As he speaks, Mrs. Hardy stands up and starts to take off her jacket. He cuts himself off to say

Dr. K: If you will
undress— put the gown on with the opening in the back
and uh
just step around here.

He gets up, draws the curtain along the center of the room, and gestures her behind it with him. He takes a gown for her out of the drawer under the examination table, comes out again, and sits at his desk with his back to her and works on his notes.

In the absence of explicit instructions about the removal, retention, and replacement of evidential boundaries, persons can be puzzled about how complete their transformation is meant to be. From behind the curtain Mrs. Hardy asks,

Mrs. H: Take the shoes and stocking off?
Dr. K: Uh
Do you have panty hose on?=
Mrs. H: Yeah
Dr. K: Yeah then take them=
Mrs. H: all off.
((After a moment))
Dr. K: How're you doing all right?
Mrs. H: Umhm.

Dr. K: Come let's pull the curtain back.
Sit down over here.

She starts to emerge from behind the curtain, he draws it open, she goes back and sits on the end of the examination table.

With the change of clothes, the person's transformation into a patient could be supposed to be complete. Dr. Kleinfeld approaches Mrs. Hardy and, without comment, picks up her right forearm, tucks it under his right elbow, wraps the blood-pressure cuff around her upper arm and pumps up the device, holding her elbow with his left hand, fingers on the crook, thumb underneath. Compliant with these cues, she attempts to sustain the position into which Dr. Kleinfeld has put her body, and he says to her, "O.K., let your arm down now. Let your arm down, that's good." The physician's arrangements to change realms take the form of instructions to the person to handle her or his own body as an object. At the close of history-taking, Dr. Silverberg and Alex are still discussing Dr. Malinowski's hoarse voice.

Dr. M: I don't know why I have it maybe because I'm putting on the show.
Dr. S: I would like to examine you.
Dr. M: For this I came.
Dr. S: I will lead you into the examining room?
Dr. M: All right.
Dr. S: I would like you to
take everything off
down to your undershorts.
And have a seat on the table.
O.K.?=
Dr. M: Do you mean when— over there.
Dr. S: We're going
right next door.

Dr. Silverberg opens his office door, escorts his patient down the hall, opens the door to another room, and gestures Dr. Malinowski in:

Dr. S: Right in here.
Just fine.
Dr. M: To leave the shorts and undershirts?

Dr. S: Just the shorts.
O.K.?
Be in in a second.
Dr. M: Take off the hearing-aid?
Dr. S: Leave the hearing-aid on.
Dr. M: Leave in the ears.

Dr. Silverberg goes out, closing the door, goes back to his office and returns in three minutes, knocks three times on the door, listens for a response, opens it and enters, and observes Dr. Malinowski lying down on the examination table in his shorts, and says, "Perfect." "Perfect" is an evaluative remark that comments on the arrangement of the body as an object in space. The patient's complicity in transforming his body into an object is encapsulated in Dr. Malinowski's offer to take off his hearing-aid, that is, to reduce his own auditory apprehension to the status of an object's.

The physical examination dislodges the self from the body so the body can be handled as an object. This objectification is accomplished in part by the physical disarticulation of the body into its parts. Dr. Silverberg begins his examination of Dr. Malinowski's body with his blood pressure, moves to his head, ears, eyes, nose, mouth, throat, round to his back, then his chest and heart, down to his abdomen, genitals, legs, up to the arms, fingers, ears again, and finally the prostate and anus.

In women patients, the use of drapes during the physical examination enhances the disarticulation of the body into parts, circumscribing each part as a separate object of scrutiny. Mrs. Hardy has on her gown with the opening at the back. Dr. Kleinfeld asks her to lie down on the examination table and lays a paper drape across her lower abdomen like an apron. He then lifts her gown up from underneath the drape and holds it crumpled just under her breasts so that only her belly is exposed. He palpates her belly with his right hand, releases her gown, which slightly unfolds, and continues the abdominal examination with both hands. With male patients a different delicacy is observed. Dr. Malinowski is wearing loose white cotton trunks that button at the top. Dr. Silverberg asks him to lie down on the examination table, listens to his heart, palpates his abdomen, working from the upper aspect downward. Then he unbuttons the patient's shorts in the front, opens the edges and folds them back, examines his genitals, and folds the shorts together again, leaving them unbuttoned. This service permits the physician to examine the patient's genitals without making the patient expose himself to the physician.

The Body as Self

Despite the reframing of the body as object, its transformation is always incomplete. It is for that reason that metonymic reference by physicians to patients as parts of their bodies as diseases ("The kidney in 101" or "How's the coronary") are concealed from persons. They show too little attention to bodies as selves. Physicians must also, however, be concerned not to show too much attention to bodies as selves. Some parts of the body are incorrigibly symbolic, especially the sexual parts. Because sexual attentions to the body can also involve its disarticulation into parts, the examination of breasts, genitals, or buttocks is hedged with further evidential boundaries. Here drapes work not only to circumscribe some parts of the body but also to conceal others. Dr. Kleinfeld says to Mrs. Hardy, "O.K., I want you to roll over on your side toward the wall. You can bend your knees up to your chest." She turns away from him. He tucks her gown under her side at the back and folds the drape over her legs so that only the fold of her buttocks shows. He puts on gloves, tells her to relax and take a deep breath, leans down, resting his left hand lightly on her upper buttock, and inserts his index finger into her anus. There are, of course, other positions that can be used for rectal examinations, the classic one being the supine, legs-up-and-apart position used by gynecologists. This one, unlike that, excludes the possibility of eye contact, and (in both males and females) partially conceals the genitals. Here drapes, gloves, eye aversion, and leg position interpose evidential boundaries between the physician and his perception of the patient. The intent of these arrangements appears to be to ensure that this, the touchiest of transactions, can transpire wholly within the reframe; the patient wholly object, silent and passive; the physician wholly operator, concentrated and active. But it never does. Even here the physician attends to the presence of a self in the body. As he palpates the interior of her rectum, Dr. Kleinfeld says, "Oh, I know that's not comfortable." And Mrs. Hardy acknowledges her inherence in her body with a wince and a thin whine.

The incompleteness of the transformation of the body from subject to object is not an imperfection but an intention. In the course of the physical examination, Dr. Silverberg says to Dr. Malinowski, "I'd like to examine your prostate and your anus," and asks him to roll over on his side and bring his knees up to his chest. The patient works his shorts down over his hips as he turns over, and Dr. Silverberg helps him tuck them down under his buttocks.

Dr. S: I will be very gentle.
Dr. M: You will be gentle?
Dr. S: Absolutely.

The doctor puts on his gloves and says,

Dr. S: Bring your knees
up here too.
Good.
O.K.
Fine.
Bring your knees up.
//
Like this. *((Lifts the professor's knees.))*
That's good.
Just breathe in and out
and relax as best you can.
This
will be uncomfortable.

He kneels, inserts his finger briefly and withdraws it.

Dr. S: Finished.
Looks fine.
Dr. M: I should have them back?
Dr. S: Yes please.
((And Dr. Malinowski pulls up his shorts.))

THE ETIQUETTE OF TOUCH

The dislodgment of the self from the body is designed to preserve the social persona from the trespasses of the examination. But the self is so deeply worked into the body that physicians must also be concerned to preserve the dignity of the self, the social person whose lodgment happens to be the body. This dual attention to the body as incarnate and discarnate, self and object, is handled by a delicate manipulation of frames and boundaries that might be called an etiquette of touch. In response, patients can cede their bodies to the realm of medicine by abstracting themselves for the nonce. But they can also, equally delicately, compose themselves outside of their bodies, for instance, in the realm of narrative. Either gesture of disembodiment preserves the etiquette of touch.

NARRATIVE EMBODIMENTS

Persons are tender of their bodies as if their selves inhered in its organs, vessels, tissues, bones, and blood; as if they were embodied. For us, the body is the locus of the self, indistinguishable from it and expressive of it. I experience myself as embodied, incorporated, incarnated in my body. To be present in the flesh is to evidence this implication of my self in my body.

Medical examinations threaten the embodied self with untoward intimacies. The accoutrements of propriety are stripped away: I appear in nothing but my body. What follows has the structure of a transgression, an infringement, but one in which I am complicit. I disclose my body to the Other, the stranger, the physician. John Berger writes, "We give the doctor access to our bodies. Apart from the doctor, we only grant such access voluntarily to lovers—and many are frightened to do even this. Yet the doctor is a comparative stranger" (Berger and Mohr 1976, 64).[7] To deflect this threat to the embodied self, medicine constitutes itself a separate realm in which the body as lodgment of the self is transformed into the body as object of scrutiny: persons become patients.

This transformation is intended to protect the sensibilities of the social self from the trespasses of the examination. Whatever the medical business of the examination, its phenomenological business is to displace the self from the body. However, people can perceive rendering the body an object as depersonalizing, dehumanizing, or otherwise slighting to the self. This sense of dehumanization is well attested to in both popular and social scientific literature. Elliot Mishler locates dehumanization in the discourse of medicine, where he describes it as the conflict between the voice of medicine, which is understood to dominate during medical examinations, and the voice of the lifeworld, which is suppressed in a way, he argues, that leads to an "objectification of the patient, to a stripping away of the life-world contexts of patient problems" (1984, 128). The disparity between the physician's intention and the patient's perception establishes the context for "gaps," "distortions," and "misunderstandings" between patients and physicians (Mishler 1984, 171).

Because of their sense of the loss of self—a well-founded sense if also a well-intentioned loss—patients can have some impulse to reconstitute a self during medical examinations. This reconstitution can be undertaken by the patient in one of two moves: either by breaking the framework of the realm of medicine by disattending, misunderstanding, or flouting its conventions—Mishler describes this as interrupting the voice of medicine with the voice of the lifeworld (1984, 63)[8]—or by maintaining the framework but inserting into the realm of medicine an enclave of another ontological status, specifically, a narrative enclave.

Rules for producing narratives on ordinary occasions require that they be set off by their frames from the discourses in which they are embedded (see Young 1986, 277–315). Narrative frames—prefaces, openings and beginnings, endings, closings and codas—create an enclosure for stories within medical discourse. The discourse within the frames is understood to be of a different ontological status from the discourse without. In particular, the storyrealm, the realm of narrative discourse, conjures up another realm of events, or taleworld, in which the events the story recounts are understood to transpire (Young 1986, 15–18). It is in this alternate reality that the patient can reappear as a person. In making such an appearance, the patient becomes, as Natanson puts it, "the self constructing for itself the shape of the world it then finds and acts in" (1970, 23). This possibility arises among what Alfred Schutz calls "multiple realities" (1973, 245–262), different realms of being, each with its own "metaphysical constants" (Schutz in Natanson 1970, 198), which individuals conjure up and enter into by turning their attention to them.

Embodying the self in a narrative enclave respects the conventions of the realm of medicine and at the same time manages the presentation of a self, but of one who is sealed inside a story. An inverse relationship develops between the uniquely constituted narrative enclosure in which a patient presents a self and the jointly constituted enclosing realm in which the patient undergoes a loss of self. Stories become enclaves of self over the course of an occasion on which the body inhabits the realm of medicine.

Goffman argues that persons are in the way of presenting themselves, guiding controlled impressions, not necessarily to deceive, but rather to sustain a reality, an event, a self. Structurally, the self is divided into two aspects, the performer who fabricates these impressions and the character who is the impression fabricated by an ongoing performance which entails them both (Goffman 1959, 252). On ordinary occasions, then, persons do not provide information to recipients so much as present dramas to an audience (Goffman 1974, 508). It is here that the theatrical metaphor for which Goffman is famous takes hold: talk about the self is not so far removed from enactment. We do not have behaviors and descriptions of them but a modulation from embodied to disembodied performances. Storytelling is a special instance of the social construction of the self in which "what the individual presents is not himself but a story containing a protagonist who may happen also to be himself" (Goffman 1974, 541). Here the performer in the realm of medicine fabricates a self in the realm of narrative. In this instance, embodying the self in stories occurs in circumstances in which the self is being disembodied, a complication of the matter Goffman has called "multiple selfing," that is, the evolving or exuding of a second self or several selves over the course of an occasion on which the self is being presented (1974, 521n).

This narrative embodiment of the person moves against the patient's progressive disembodiment over the course of the examination. Medical examinations are divided into two parts: the history-taking and the physical examination. These internal constituents of the realm of medicine are bounded by greetings and farewells that mark the transition between the realm of the ordinary and the realm of medicine. The shift from greetings, in which the physician emerges from his professional role to speak to his patient as a social person, to history-taking, in which the physician elicits information from the patient about his body, is the first move toward dislodging the self from the body. The patient's social person is set aside to attend to his physical body.

In the course of his medical examination, Michael Malinowski, the professor of Jewish literature and history, tells three stories in which he appears as a character. The links and splits between the realm of medicine and the realm of narrative illuminate the nature of narrative, the nature of medicine, and the nature of the self. After Dr. Silverberg shakes hands with the professor and his son in the waiting room, he escorts them to his office, and there begins to take the patient's history. The shift from the waiting room to the office reifies the transition between realms. The history-taking reorients the person's attitude toward his body in two respects: it invites him to regard his body from outside instead of from inside, and it invites him to see it in parts instead of as a whole. Dr. Silverberg's inquiries direct the patient to attend to his body as an object with its own vicissitudes, which he recounts with the detachment of an outsider. In so doing, Dr. Malinowski suffers a slight estrangement from his own body. In making these enquiries, Dr. Silverberg asks about the parts of the body separately, disarticulating it into segments. So Dr. Malinowski's body undergoes a fragmentation. Since the self is felt to inhere in the body as a whole and from the inside, these shifts of perspective tend to separate the self from the body. It is against the thrust of this ongoing estrangement and fragmentation that the professor sets his first story, which might be thought of as the story of the liberation. Dr. Silverberg has shifted from general inquiries about the whole body—height, weight, age, health—to specific inquiries about the eyes, the throat, and the blood. He then continues.

Story 1: The Liberation

Dr. S: Have you ever had any problem with your heart.
Dr. M: No.
Dr. S: No heart attacks?
Dr. M: Pardon me?
Dr. S: Heart attacks?
Dr. M: No.
Dr. S: No pain in the chest?
Dr. M: No pain in the chest.
Dr. S: I
noticed that=
Dr. M: I am a graduate from Auschwitz.
Dr. S: I know— I heard already=
Dr. M: Yeah.
I went there when— I tell Dr. Young about this

and
after Auschwitz
I went through a lot of— I lost this
Dr. S: Umhm.
Dr. M: top finger there
and
I was in a—
after the liberation we were under supervision of
American doctors.
Dr. S: Yeah?
Dr. M: American doctors.
Dr. S: Right.
Dr. M: And it uh
I was sick of course after two years in Auschwitz I was quite uh
uh exhausted.
And later I went through
medical examination
in the American Consul
in Munich
Dr. S: Yeah?
Dr. M: and I came to the United States.
Dr. S: Right?
Dr. M: In nineteen hundred forty-seven.
Nineteen forty-six—
about nineteen forty-seven.
One day—
I lived on Fairfield Avenue
I started to spit
blood.
Dr. S: Right?
Dr. M: Yeah?
And I called the doctor
and he found that something here *((gestures to his chest))*
Dr. S: Tuberculosis?
Dr. M: Somethin— yeah.
And I was in the Deborah
Sanitorium for a year.

Dr. S: In nineteen forty-seven.
Dr. M: I would say forty-seven and about
month of forty-eight.
. . .
Dr. S: Back
to your heart.

This story conjures up a taleworld, the realm of Auschwitz, which is juxtaposed to the ongoing history-taking. The preface, "I am a graduate from Auschwitz," opens onto the other realm. Prefaces are a conventional way of eliciting permission to take an extended turn at talk in order to tell a story (H. Sacks 1992, II, 40). In response to what he perceives as a digression from the realm of medicine, Dr. Silverberg says, "I know—I heard already." Having heard a story is grounds for refusing permission to tell it again (Goffman 1974, 508). Dr. Malinowski persists in spite of this refusal, thus overriding one of the devices available to physicians for controlling the course of an examination, namely, a relevancy rule: that the discourse stay within the realm of medicine. To insert the realm of narrative into the realm of medicine, the professor initially breaks its frame. But in so doing, he substitutes another relevancy rule: topical continuity. Like the history-taking, the taleworld focuses on a part of the body, the chest. It is this part of the body that the professor uses to produce topical continuity between the history-taking and the story. However, it is not the chest but the heart on which the physician is focusing. When he returns talk to the realm of medicine with the remark, "Back to your heart," he is at the same time protesting the irrelevance of this excursion. As is apparent here, the rule for topical continuity, the selection of a next discourse event which shares at least one element with a previous discourse event, permits trivial connections between discourses and, by extension, between realms. But there is a deeper continuity here. Both the realm of Auschwitz and the realm of medicine address the body.

Mishler notes that the struggle between voices for control is associated with disruption of the flow of discourse (1984, 91). He writes that to see departures from the medical paradigm as interruptions is to privilege the physician's perspective (1984, 97). "I am proposing an interpretation of the medical interview as a situation of conflict between two ways of constructing meaning. Moreover, I am also proposing that the physician's effort to impose a technocratic consciousness, to dominate the voice of the lifeworld by the voice of medicine, seriously impairs and distorts essential requirements for mutual dialogue and human interaction. To the extent that clinical practice is realized through this type of discourse, the possibility of more humane treatment in medicine is severely limited" (1984, 127). This is so despite the fact that the objectification of the body can be intended to protect the sensibilities of the person. On the other hand, to see the dominance of the medical paradigm as an imposition is to privilege the patient's perspective. My concern here is the rhythm of interplay between perspectives, discourses, or realms.

In the realm of medicine, the dismantling of the body continues with Dr. Silverberg's inquiries about the heart, breath, ankles, and back; he recurs to whole body concerns with inquiries about allergies, habits, and relatives; then he goes on to segment the body into the skin, head, eyes again, nose, throat again, excretory organs, stomach again, muscles, bones, and joints. Into this discourse, the professor inserts his second story, the story of the torture. This story is also about a part of the body, the finger, and so again maintains a parallel with the realm in which it is embedded, although not the strict tie of topical continuity. Having created an enclosure in medical discourse for the Auschwitz stories earlier on, Dr. Malinowski now feels entitled to extend or elaborate that taleworld (Young 1986, 80–99). This story is tied, not to the discourse that preceded it, but to the previous story in which he mentions his finger. As if in acknowledgment of the establishment of this enclosure, Dr. Malinowski's preface, "I was not sick except this finger," elicits an invitation from Dr. Silverberg to tell the story: "What happened to that finger." The taleworld is becoming a realm of its own.

Story 2: The Torture

Dr. M: No.
I don't know.
I tell you—I told you Dr. *((to me))* I don't
during the twenty-three months in Auschwitz
Dr. S: Yeah?
Dr. M: I was not sick except this finger.
Dr. S: What happened to that finger.

Dr. M: I wa—
I tell Dr. Young
I was sitting
((coughs)) you have something to drink
Dr. S: Yeah.
I have for you.
Dr. M: Yeah.
I was sitting at the press—
the machine
I don't know how to say in English
—a machine or
[[[ ]
Dr. S: I understand.
Dr. M: Anyway I had to put in this was
iron
and I had to put in— in here with the right hand to put this which made a hole or whatever it did.
Dr. S: Made a hole in your finger.
Dr. M: No.
Made a hole here *((in the piece of iron)).*
My finger got it.
And behind me was an SS man.
The SS was walking.
And he stood behind me
and at one moment he pushed me.
Just— this was a— a— a—
daily sport.
And instead to put the iron in I put my finger in.
/
But otherwise I wasn't sick.

The shift from taking the history to giving the physical examination involves moving to another space, the examining room, which is an even more narrowly medical realm. Dr. Silverberg then takes his patient to the examining room down the hall and leaves him to take off his clothes. Clothes are the insignia of the social self. Their removal separates the body from its social accoutrements. This reduction of the social self along with the enhancement of the medical realm completes the dislodgment of the self. What remains is the dispirited, unpersoned, or dehumanized body.

During the physical examination, the body is handled as an object. When Dr. Silverberg returns, he picks the patient's right hand up off his chest, holds it in his right hand, and feels the pulse with his left fingertips. Here is the inversion of the initial handshake which enacted a symmetry between social selves; the physician touches the patient's hand as if it were inanimate. The examination is the rendering in a physical medium of the estrangement of the self and the fragmentation of the body. The person's internal perspective on his own body is subsumed under the physician's external perspective and the whole is disarticulated into parts. Of course, there is still talk: questions, comments, instructions; but now such remarks are inserted into interstices between the acts, the investigations, the physical manipulations that structure the examination. Henceforth, for the course of the physical examination, the patient's body is touched, lifted, probed, turned, bent, tapped, disarranged, and recomposed by the physician. It is here that the absence of the self from the body can be intended as a protection: the social self is thereby preserved from the trespasses of the examination. These are committed only on an object.

After completing the physical examination, including the genital examination, Dr. Silverberg has Dr. Malinowski sit up and looks again at his arms and hands. On the patient's forearm, the faint tattoo of his concentration camp number is visible. At this point, Dr. Silverberg asks the patient to touch his nose with the tips of his fingers and as he does so the patient alludes to a bump on his skull: "I have to tell you how I got that." And the physician responds, "How." Despite this invitation, Dr. Malinowski appears uncertain about the propriety of inserting a story into this most objectified realm.

Story 3: The Capture

Dr. M: I have to tell you how I got that *((the bump)).*
Dr. S: How.
Dr. M: Should I talk here?
Dr. S: You=
Dr. M: Can I talk here?

Dr. S: Sure.
Dr. M: You already know *((to me)).*
When I (s— try) to go to the border
between Poland and Germany
Dr. S: Yeah?
Dr. M: I wanted to escape
to the border over Switzerland=
Dr. S: Umhm.
Dr. M: as a Gentile.
Dr. S: Yeah?
Dr. M: When they caught me
they wanted investigation
/
Dr. S: That it?
Dr. M: (At)
the table was (sitting) near me
and (his arm) was extending behind me
with— how the police ha— how do you call it.
A police club?
Dr. S: Nightstick.
Dr. M: Nightstick.
Dr. S: Umhm.
Dr. M: And they—
I had to count
and they hit me twenty
times over the head.
And er— he told me *zahlen*
—*zahle* means you count.
And after the war—
after the liberation shortly about two three days
American Jewish doctors came
they (examined us)
and he told me
that I have
a nerve splint here?
Dr. S: Yeah.
Dr. M: And this made me be deaf.

The physician then examines the patient's ears, and finally his prostate and rectum. So here, suspended between the genital and rectal examinations, the two procedures toward which the displacement of the self from the body are primarily oriented, is the professor's third and last story. Once again, the story is about a part of the body, the ears, which maintains a continuity with the realm of medicine. But it is also about another part of the body, the genitals. As he mentions, Dr. Malinowski has already told me this story when I talked to him in the waiting room to get his permission to observe and tape-record his examination. He told me that he and a friend had decided, boldly, to cross the border out of Poland into Germany and work their way across Germany to the Swiss border. They carried forged papers. He himself got through the border and was already on the other side when something about his friend aroused the border guard's suspicion and they called him back. To check their suspicions, the guards pulled down his pants and exposed his genitals. Jews were circumcised. This story is concealed as a subtext directed to me within a text directed to the physician. On this understanding, the positioning of the story between the genital and rectal examinations has a tighter topical continuity than is apparent on the surface.

NARRATIVE EMBODIMENTS

Stories are sealed off from the occasions on which they occur, here from the realm of medicine, as events of a different ontological status. For that reason they can be used to reinsert into medicine an alternate reality in which the patient can reappear in his own person without disrupting the ontological conditions of the realm of medicine. Stories about the realm in which Michael Malinowski appears as a character, the world of Auschwitz, might be supposed to be inherently theatrical, on the order of high tragedy. But the boundary between realms insulates medicine in some measure from the tragic passion. The apertures along the boundary through which the realms are connected are here restricted to parts of the body. In telling these stories, Dr. Malinowski is not merely playing on his hearers' emotions. He is rather reconstituting for them the ontological conditions of his world and, having done so, inserting himself into that realm as a fully embodied person. Besides creating a separate reality, telling stories during a medical examination creates a continuity between the two realms that converts the ontological conditions of the realm of medicine precisely along the dimension of the body.

The stories are tokens of the man, talismans of the salient and defining history which has shaped him. They are not, on that account, unique to this occasion, but are invoked as touchstones of his presence (as they were, for instance, for me when we talked before the examination). They present a person whose life is wrought around an event of existential proportions. Auschwitz was a life-pivoting, world-splitting event: time is reckoned before-Auschwitz and after-Auschwitz; space is divided by it. Not only has he lost a country, a language, and a childhood, but he has also lost a life form. Before Auschwitz, he had a wife and child in Poland; the son who has brought him today is the only child of a second marriage made in the United States after the war. Dr. Malinowski mentions once that he had two sisters: one "perished"; the other died a few years ago of cancer.

The sequential order of events in a story replicates the temporal unfolding of events in the realm it represents (Labov 1972, 359–60). This replication is supposed by some social scientists to extend to the sets of stories which are strung together to make a life history. In this instance, the sequential order in which these stories are told does not replicate the temporal order in which the events they recount occurred. Dr. Malinowski tells about the liberation first, then the torture, and finally the capture. There are of course clear contextual reasons for this that have been detailed here in terms of topical continuity. But I would like to suggest a deeper reason for their array. These stories cluster around Michael Malinowski's sense of self. Auschwitz provides what I would like to call centration: life is anchored here, everything else unfolds around this. The set of stories that make up the Auschwitz experience could be told in any order. There is an implication here for the use of narrativity in the social sciences. By insisting either on the notion that temporally ordered events are presented sequentially among as well as within stories or on re-ordering stories to present them so, social scientists have misunderstood the shape of experience: a life is not always grasped as a linear pattern. Serious attention to narrativity in what John Shotter and Kenneth Gergen call "texts of identity" (1989) will not force the sense of self into the pattern of narrative but will instead deploy narrative to discover the sense of self.

In so presenting the man and reconstituting the ontological conditions of his world, these stories attain the status of moral fables and lend the medical examination a delineation that renders the etiquette of touch an ethical condition. Not that the stories are warnings to the physician against similar transgressions. On the contrary, in the existential context of these stories, what might otherwise be seen as indignities to the body are transmuted into honors: the physician is a man whose touch preserves just those proprieties of the body that are infringed at Auschwitz.

The body in the taleworld is the analogue of the body in the realm of the medical examination, connected to it part for part, but inverted. The stories spin out existential situations in which the self is constrained to the body. In the first story, "The Liberation," the part of the body is the chest and the mode of insertion of the self in the body is sickness. In that condition, the self cannot transcend its absorption in its bodily discomforts; its sensibilities are sealed in its skin. Drew Leder suggests: "The body is thematized at times of dysfunction or problematic operation" (1990, 85).

In the second story, "The Torture," the part of the body is the finger and the mode of insertion of the self in the body is pain. The self is jolted into the body, its sensibilities concentrated in its minutest part, the tip of a finger. Elaine Scarry argues that the self as the essential center of my experience and as my embodied bridge to the world are split by torture. "The goal of the torturer is to make the one, the body, emphatically and crushingly *present,* by destroying it, and to make the other, the voice, [which embodies my extension into the world], *absent* by destroying it" (1987, 49). "Pain," Leder expatiates, "exerts a power that reverberates throughout the phenomenological field, shifting our relations both to the world and to ourselves. There is a disruption of intentional linkages and a constriction of our spatiality and temporality to their embodied center" (1990, 79). He continues, "disease, like pain, effects a disruption of intentional links and a spatiotemporal constriction" (1990, 80). And "this intentional disruption and spatiotemporal constriction correlates with a heightened thematization of the body" (1990, 81).

In the third story, "The Capture," the parts of the body are the head and the genitals, and the mode of insertion is humiliation. Here the body is emblematic of the man, literally inscribed with his identity. Its degradations are his. Injuring the body changes the circumcision from an inscription of honor into a source of torment, obliging him either to disown his body and betray his honor or to own his body and accept his torture. The medical examination, by contrast, can be understood to release the self from the body.

The phenomenological cast of the taleworld in which the self is implicated in the body is set against the phenomenological cast of the realm of the medical examination in which the self is extricated from the body. The medical history of the tuberculosis, the severed fingertip, the deafness, which could be detached from their etiology is instead enfolded in the personal history of the concentration camp and recounted as a story. So Auschwitz is invoked not as the cause of these dissolutions of the flesh, but as the frame in terms of which we are to understand what has befallen the body and, it transpires, the frame in terms of which we are to understand what has become of the man. To see the fact that both the realm of medicine and the realm of narrative are about the body as their essential connection is a trivial rendering. The stories are transforms of the ontological problem that is central to the examinations: the fragile, stubborn, precarious, insistent insertion of a self in the body.

TWO

• • •

PERCEPTUAL MODALITIES • GYNECOLOGY

The living body is more than a thing extended in visual space. It is first and foremost the center of a tactile-kinesthetic world that, unlike the visual world, rubs up directly against things outside it and reverberates directly with their sense. The tactile-kinesthetic body is a body that is always in touch, always responding with an intimate and immediate knowledge of the world about it. Reduced in status to a visual object, the body loses this quintessential sensorium.

—*Maxine Sheets-Johnstone*

Medical discourse appropriates the body as a site of representation. There, on my skin, in my viscera, are inscribed the conceits of medicine. These conceits, exquisitely schematized, are figured in anatomical drawings, anatomical texts, and anatomical practices from whence they are read back into the body. Anatomy is adduced to support a politics of the body. Not only do anatomical representations shift in order to substantiate different arguments but also the anatomy itself transmutes in the eye of even the most scrupulous anatomist: witness Leonardo da Vinci's pellucid sketch of the ovarian vein, a vessel putatively connecting the breasts to the uterus. By virtue of this pathway, according to the sixteenth-century physician Laurent Joubert, the bodily fluid that issues as blood from the womb can be transformed into milk in the breasts. "And is it not the same blood, which, having been in the womb, is now in the breasts, whitened by the vital spirit through its natural warmth?" (Joubert 1579, 451). Barbara Duden writes that up to the eighteenth century, "the fluid inside the body could apparently assume different forms, yet remain always the same substance. The inside was a porous place, a place of metamorphosis: fluids changed in the body, they transformed their materiality, form, color, consistency, and place of exit, and yet apparently they remained essentially alike" (1991, 109). Differences between bodies were matters of degree. In consequence of their more intense body heat, for instance, men not only kept their humoral fluids purer than women but also expulsed the genital organs women retained inside. Thomas Laqueur juxtaposes Renaissance anatomical drawings which represent females as inversions of males, and vice versa, with eighteenth-century representations of gendered anatomies as different in kind, at once constructing and reflecting the shift from the ancient "metaphysics of hierarchy" to an "anatomy and physiology of incommensurability" (1990, 6).

THE DISCURSIVE BODY IN MEDICINE

On the modern body, the conceits of medicine move to reconfigure a social person as a medical object. I am translated bodily from the realm of the ordinary to the realm of medicine where I take on the insignia of that other discourse. My corporeal self becomes a medical body. The understandings constitutive of medical discourse are thus instantiated bodily; a bodylore sports the lineaments of a metaphysics.

Not only is the discourse of medicine inscribed on the body but also the body substantiates the discourse. Through the body's perceptual modalities, I constitute medicine a realm of experience. I am invested bodily in a world even as I invent it, "the self constructing for itself the world it then finds and acts in" (Natanson 1970, 23). I apprehend its properties corporeally. Medicine is, for me, a visual, auditory, tactile, olfactory, gustatory reality. I do not so much perceive a reality as produce it. Reality issues in, as, and from my body as part of its corporeal nexus. There is, Richard Zaner writes, "no way to section off the body from itself, or the embodied self from its circum-standing world" (1981, 98). The body is what Zaner calls "contextured."

> Throughout the spectrum of bodily gestures—embracing, holding off, handshaking and fist-waving, nodding and kneeling, yielding and confronting—there simply is no way to separate out an "inside" which (by

> magical means only) would then get "pressed" to the outside. The weeping of a small child, the blushing of an embarrassed person, the sighing of relief—all are gestural displays saliently presenced within specific circumstances which always include and comprehend other embodied beings, other embodied human persons, as well as what comes to be differentiated out as physical entities. *Embodiment, thus, is itself a gestural contexture within a still wider context:* the world of concretely environing other live beings, other persons, social life, and that of "nature." (Zaner 1981, 66, emphasis in original)

Corporeality includes the self, the body, and the world. I am, Zaner writes, "fundamentally locused within and by my central body and my surrounding zonal sphere of immediate bodily action and spatiality. *My embodying organism is my complexly contextured 'habitus'* . . ." (1981, 98). The way I attest perceptually to the realm I inhabit, the way I lend myself to my reality, makes it possible to mold my production of a reality by constraining my perceptual modalities.

PRODUCING PERCEPTUAL REALITIES

Perceptual modalities, preeminently vision, can lift me away from the source of perception on my body toward its perceptual object. I am aware of the trees outside, the cabinet in my room, the writing on the page, but not of myself seeing. A disruption of the visual field, the sun hurting my eyes, turns my awareness to the source of perception on my body rather than its object elsewhere. Hearing, too, can extend me out of my body so that I am aware of sounds as elsewhere. Again, a sound that hurts my ears shifts my awareness from the world to the site of hearing on my body.

In visual and auditory perception, the body "stands out" from itself, near its objects. It is what Drew Leder (1990, 21) calls "ec-static." As I am engaged by objects of my perception, my body withdraws, recedes, "disappearing from self-awareness" (1990, 26). This Leder calls the recessive body. The ecstatic body and the recessive body are modes of "absence," "ways the body can *be away* from itself" (1990, 26). At the least, in the act of perception, the body undergoes "focal disappearance" of the perceptual apparatus on the surface of the body (1990, 53). I do not attend to my eyes seeing or my ears hearing. "Directed ecstatically outward, my organs of perception and motility are themselves transparent at the moment of use" (1990, 53). Typically, I am with my objects of perception and without my body.

> It is thus possible to state a general principle: insofar as I perceive through an organ, it necessarily recedes from the perceptual field it discloses. I do not smell my nasal tissue, hear my ear, or taste my taste buds but perceive with and through such organs . . . As such, one's corporeality recedes more thoroughly than in touch, whose reciprocity and feeling of impact calls one back to the copresence of the body with its object. In taste, the experience of world and body is perhaps most closely interwoven; the act of perceiving involves the literal incorporation of the perceived. (1990, 14–15)

Touch inmixes the perceptual experience of the object and the source of perception. I am aware by turns of what I feel and of myself feeling. The boundary between skin and thing is blurred, ambiguous, shifting. Olfaction erases the boundary between body and object almost altogether. I find it difficult to distinguish the sharp tang of salivation from the scent that provokes it. Taste actually incorporates its object. Here, too, inimical sensations assault and bring forward the body. Nasty tastes, noxious smells, painful touches "thematize" the body in my awareness, to adopt the phenomenological usage.

Each sensory receptor sustains its object at a characteristic distance from the body. Touch, like taste and smell which are forms of touch, requires propinquity. Sound holds its objects at an astonishing remove from the body; sight even further. Indeed, as Michael Jackson points out, visuality "implies a *spatialization* of consciousness in which knower and known are located at several removes from one another and are regarded as essentially unalike, the one an impartial spectator, the other subject to his gaze" (1989, 6). Visual and, to a certain extent, auditory apprehensions estrange subject and object. They are the perceptual modalities that underpin objectivity in medical discourse. Tactile apprehensions, in their various modulations, draw the object into the body. This perceptual modality threatens objectivity in medical discourse.

Perturbations of perception, typically pain, return my attention to my eyes and ears, my nose and tongue, my skin. But I can also recover awareness of these perceptual modalities by turning my attention to them. My surface remains marginal to awareness; it can be recalled, thematized in consciousness. My interior is more radically recessive. The viscera are in a persistent state of "corporeal disappearance" (Leder 1990, 36). Indeed, it is what Leder calls the "perceptual reticence of our viscera" that obliges physicians to probe us, to slip into slits made or found on the body. As a consequence of this depth disappearance, we are conducted to our own viscera as to an alien terrain.

((The gynecologist, David Casaccio, describes to Roberta Lubovitch a photograph of her own cervix.))

Dr. C: You want to take a look at this?
You can look up in— into the mouth of your womb
And see this little tissue right here?

Ms. L: Mmhm.

Dr. C: See the red and the pink?
That's the border
that we want to see.
And you can see it all the way around.
I biopsied a little bit here
and then underneath this speculum I biopsied
and that's the only two biopsies that I did.
Actually the cervix looks pretty normal.

(April 29, 1987, Lefkowitz transcript, 12–13)[1]

But visceral sensations are not altogether inarticulate. Unlike exteroception, the perception of external objects, which I experience without "gaps or crevices," interoception has "ineluctable discontinuities" (Leder 1990, 42). Visceral sensations are "vaguely situated with indistinct borders" (Leder 1990, 41) rather than sharply focused at their sites of origin. Unlike perceptual apparatuses on the surface of the body, they do not seem to reveal themselves along various modalities but rather are "experienced as modulating a single dimension of perception" (1990, 40). These sensations cannot be recalled or brought to awareness but may, in dysfunctional states, announce themselves, or what Leder calls "dys-appear" (1990, 85). Hence, the metaphysical contrasts Leder investigates between absence and presence, the recessive body and the ecstatic body, disappearance, that is, absence, and dys-appearance, that is, forced presence, aspects of what Jacques Derrida calls "the metaphysics of presence" (1973, 101).

Pain obtrudes my body on my awareness. It retracts me from my extensions out attentionally into space and backward memorially or forward projectively in time, and injects me into my body. I am obliged to act "toward" rather than "from" my body (Leder 1990, 79). Interoception precludes exteroception (Leder 1990, 39). Extreme pain, as Elaine Scarry puts it, unmakes the world (1987, 20). I am trapped in the "discrepancy between an increasingly palpable body and an increasingly substanceless world" (Scarry 1987, 30). "Rather than enabling the person's movement out of the boundaries of the body into the shared realm of extension, [extreme pain] instead brings about a continual contraction and collapse of the contents of world-consciousness . . . it deconstructs already existing objects in order to inflict pain" (1987, 144–145). I am so riven to my body that everything else vanishes.

Pain not only unmakes the world, according to Scarry, it also unmakes language. "Physical pain is not only itself resistant to language but also actively destroys language, deconstructing it into the prelanguage of cries and groans. To hear those cries is to witness the shattering of language" (1987, 172). Thus, even as pain is inexpungibly present to its perceiver, it is irretrievably remote from its hearer (1987, 3–4). This is a matter not merely of the inexpressibility of pain but of its interiority. It is the core philosophical instance of private experience, inner sensations, other minds, of the radical inaccessibility to me of the presence of the Other in the flesh.

BREAKING THE MEDICAL BODY INTO ITS DISCOURSES

Gynecological examinations return me to my body by encroaching on my viscera. Tugs, disturbances, intimations arrive intermittently from my remote interior. I am incapable of detachment. Unlike the external perceptual modalities, "visceral sensations grip me from within, often exerting an emotional insistence" (Leder 1990, 40). At the same time, I am obliged by the conventions of medicine to reframe my own body as an object, to come upon it as if from outside, to inspect it, to allude to it, to attend to it, as to a thing. I am induced to disengage, to detach myself from my corporeality, as it were, to disembody. It is to protect the social person from what would otherwise be the improprieties of the examination that the self is displaced from the body. These intimacies are now transacted on a dispirited object. Propriety entails a strict discretion between the realm of medicine in which I am an object and the realm of the ordinary in which I am a self. The insistence of my visceral sensations threatens to disrupt the inscription of objectivity on my body.

To preserve the propriety of the body in gynecological examinations, both the realm of the ordinary and the realm of medicine are set up in the same region of space and conducted over the same span of time. These alternate realities are instantiated and sustained by the architectural organization of the examination room, the discursive practices of its inhabitants, and the disposition of the body. The space, the discourse, and the body are all cleaved in two. The boundary between them is partially reified in a material barrier: the gown and drape laid over the patient's body. Roughly, what transpires underneath these coverings lies in the realm of medicine; what transpires above them lies in the realm of the ordinary.

Realm discretion is strictest during the pelvic examination. For this, the patient is lying on her back with her legs bent double and her feet mounted in stirrups on either side of the examination table. The gynecologist sits on a low stool at the foot of the table. Between them, a drape is mounted up across the patient's abdomen and thighs, tucked along the sides of her legs, and hung over the edges of the table. The drape provides what Erving Goffman calls an "evidential boundary" (1974, 215) that intervenes between the participants in the examination and their perception of each other. Visually, the drape prevents either patient or physician from catching a glimpse of the other's face between her thighs. From the perspective of the woman, the upper frontal aspect of her body down to her knees appears as a series of mounds and hollows protruding through the horizontal blue-green surfaces of her gown and drape. From the perspective of the gynecologist, the drape frames the underside of the patient's haunches which appear as a vertical surface rimmed by the forepart of her calves and feet, and enclosing her vulva, its central fold extending downward to the anus. Woman and gynecologist visually substantiate different realms. For the woman, visuality, the perceptual mode of objectivity, substantiates the realm of the ordinary; for the gynecologist, visuality substantiates the realm of medicine.

THE DISPOSITION OF PAIN, PROPRIETY, AND PLEASURE

The visual boundary creates an enclosure in the matrix of the gynecological examination for tactile perceptions. In both pain and pleasure, touch thematizes the body. Pain constrains me to my body. As Scarry suggests, the universe contracts to my skin or my body swells to fill the universe (1987, 35). Pleasure expands my bodily self. If, Scarry writes, a woman

> experiences the intense feelings across the skin of her body not as her own body but as the intensely feelable presence of her beloved, she in each of these moments experiences the sensation of "touch" not as bodily sensations but as self-displacing, self-transforming objectification; and so far are these moments from physical pain, that if they are named as bodily occurrences at all, they will be called "pleasure," a word usually reserved either for moments of overt disembodiment or, as here, moments when acute bodily sensations are experienced as something other than one's own body. (1987, 166)

The instrumental gestures of the gynecological examination cause pain, ranging from discomfort, a sort of eeriness of the interior, to torment. Pain binds the woman to her body in the realm of medicine. To permit her to express her pain is to permit her to reconstitute herself as a subjectivity in this realm and so to invite an impropriety. To prevent this, the gynecologist never admits pain to the realm of medicine as the sign of a subjective state. Expressions of pain, either in the physical form of flinches, winces, starts, grimaces, jerks, tightening or twitching the lip, arching the neck, throwing back the head, closing the eyes, blinking, or squinting, or in the audible form of inbreaths, whimpers, sighs, gasps, moans, groans, cries, or screams are typically either suppressed by the patient or "disattended," in Erving Goffman's term (1974, 202), by the gynecologist.

As Dr. Casaccio inserts the speculum, Linda Rainer tightens her lips; when he touches the interior of her vagina, she throws her head back slightly; during his examination of her rectum she closes her eyes momentarily, twists her lip, blinks, all of these grimaces transpiring outside the gynecologist's visual field (May 6, 1987, observation). This pattern of disattention can be modified when the gynecologist instructs the patient on how to control her pain in order to keep it from intruding into the examination. As he palpates her abdomen, and without making explicit the apparent discomfort that actuates his remark, Dr. Casaccio says to Mrs. Rabinowitz, "Take a deep breath. Another breath in and out" (April 29, 1987, Lefkowitz transcript, 11). In its explicitly sympathetic form, which takes account of the presence of a subject, pain management often falls to the nurse, whose alignment shifts flexibly between gynecologist and woman over the course of the examination.

((Dr. Casaccio is preparing to insert a speculum into May Schmidt's vagina.))

Dr. C: O.K. now, you're going to feel a little pressure.

((He inserts the speculum. She winces; he withdraws the speculum. The nurse, Sylvia Queens, leans over the patient.))

Nurse: You O.K.?

Mrs. S: Just a cramp in my—

((She breaks off. Nurse Queens lifts the patient's right leg up to her own shoulder level to relieve the cramp.))

Mrs. S: *((Whispers))* Oh, that's better. Thank you.

Nurse: You're welcome.

((Gloved, Dr. Casaccio touches the vulva and reinserts the speculum.))

Dr. C: O.K., there's the cervix. Are you all right?

Mrs. S: Mmm.

Dr. C: *((To nurse.))* Let's see if you can put it down now.

((The nurse lowers the patient's leg.))

Dr. C: Has it gone away, the cramp?

Mrs. S: Yes, it has.

(April 29, 1987, Lefkowitz transcript, 16)[2]

To acknowledge pain in its course would be to acknowledge the woman's presence in her flesh. Such an incursion threatens the elaborately objectified realm of medicine.

Pleasure, however, need not be intentionally excluded from the pelvic examination for the patient; it is simply not ordinarily available in the structure of the examination. For the gynecologist, as for the woman, the transmutation of these gestures into sexual overtures is improbable. John Berger writes:

> The degree of intimacy implied by the relationship is emphasized by the concern of all medical ethics (not only ours) to make an absolute distinction between the roles of doctor and lover. It is usually assumed that this is because the doctor can see women naked and can touch them where he likes and that this may surely tempt him to make love to them. It is a crude assumption, lacking imagination. The conditions under which a doctor is likely to examine his patients are always sexually discouraging. (Berger and Mohr 1967, 68)

The imaginative act that overcomes these circumstances, on either part, is unusual. Because it inmixes the body and its objects, touch is the perceptual modality of taboo, of the loss of one's own body boundaries and the incorporation of the Other. Touch can dismantle the continent objecthood of the medical body.

Other gestures attending upon the examination, however, intimate modest pleasures. To arrive at the upended position proper to the vaginal examination, the patient is passed through a sequence of postures. When the gynecologist arrives in the examining room, the patient, wearing her paper gown open at the back, is usually perched on the side of the examination table, legs dangling over the edge. Both patient's and gynecologist's bodies are enwrapped in their medical accoutrements. Nonetheless, initial interchanges take place between social persons in an ordinary realm reconstituted on the site of the medical one. They consist of greetings, including inquiries about either or both persons' well-being and modulating to specifically medical inquiries. For the next phase of the examination, the breast examination and the external abdominal examination, the gynecologist has the patient lie down on the examination table. In doing this, he may proffer his own body to assist in her adjustments of position.[3] He may, for instance, tuck his hand under her shoulder, not so much to take her weight as to guide her body to its next holding place, to shape a hollow into which her body falls. Dr. Casaccio lifts the pillow to convey to Mia Fields that she should lie down (May 8, 1987, observation). He inscribes the direction into which, along which, her body lets itself go. The woman relinquishes not her weight but her will to him. She becomes patient.

For the most intimate phase of the physical examination, the internal vaginal examination, the patient slides down the table till her buttocks depend slightly over its edge, bends her knees and puts her feet into stirrups angled out from the lower outer corners of the table. Here, the gynecologist may draw down the patient's thighs, draw out her calves, or lift up her ankles into the stirrups. Dr. Casaccio negotiates this arrangement with Anna Rabinowitz.

Dr. C: O.K., (let me) have you— right here.
((He takes hold of her right ankle in its white sock from underneath and lifts it into the stirrup as Ms. Rabinowitz slides down the table.))
Feet down— to the— stirrup
O.K., and put this other foot here. *((He lifts over her left foot.))*
Come on down a little.
That's good.
(Don't want to) slide off ( ).
That's good right there.

(April 29, 1987, Lefkowitz transcript, 11)

During May Schmidt's examination, Dr. Casaccio places both his hands gently under her thighs to signal, direct, and assist her to slide down the examination table.

Dr. C: O.K. come on down.
Little bit more.
A little bit more.
And now lie back and then I'll have you
scoot down a little bit more
Keep coming.
((As her torso nears the foot of the table, Dr. Casaccio releases Ms. Schmidt's thighs to guide her feet into the stirrups.))
Good.
That should be enough.

(April 29, 1987, Lefkowitz transcript, 16)

He provides no real lift during this maneuver, just the sensation of support and an indication of direction as the patient changes position.

As a patient sits up after the vaginal examination, the gynecologist may press back her thighs, draw her ankles together and down or haul her up by the hand. In the course of conducting these moves, the gynecologist may give the patient a salutary pat on the top of the foot, the calf, or the knee. I have even seen an Italian gynecologist, with great panache, offer a patient a valedictory pat on the bottom. Touch tangential to the instrumental acts of examination configures pleasure, the hand shaped to the curve of the shoulder as it drops, to the underside of the thigh, lifted to the lowering calf, the out-turned ankle. These gestures afford the woman small pleasures, not by feeling her own body as the presence of the Other but by slightly releasing her body into the groove provided by the Other, not by letting herself go but by letting her body go from herself, a slight displacement, a minor transcendence. Small pleasures are juxtaposed to slight discomforts.

But what is this curiously gentlemanly gesture? The gentleman interposes his own body between the body of the lady and certain possibly obdurate or obscure objects in her world, a form of what Goffman calls "protective intercession" (1976, 72). His body makes opening forays for her into new territory. She follows. His corporeal person, the more dispensable, shields hers, buffers hers, and in the same gesture, ensconces her. These little courtesies, these putative protections, these ritual kindnesses, permit her to disattend some of the exigencies of her world and oblige her to relinquish some of her sovereignty in it. At the same time, they coddle the patient, however briefly, bespeak her preciousness, her delicacy. They offer the patient, on the part of the gynecologist, an assurance of good heart, a tenderness of the flesh, a gesture to the person embodied. And this is so to a certain extent between patient and physician regardless of their actual genders.[4]

As she has been placed, as she accepts her placement, so the woman takes on her subjection. With great gentleness and tact, the gynecologist has taken her over. Her recumbent body is yielded up for inspection. She no longer assumes control of her movements; she does not move. What movement there is comes from him. What she assumes control of is her stillness. She keeps still. She makes her body quiescent so that her responses, twinges, do not actuate her flesh as winces.

> *((Dr.Casaccio is examining a patient with a herpes outbreak. As he touches her vulva, Jen Levin throws herself away from him and protests.))*

Ms. L: Oh my word.
Dr. C: I'm just going to touch you

I'm not going to poke you.
((He takes hold of a fold of tissue and pouches it out. A cluster of blisters is visible on its surface. He picks up a needle.))
Don't jump though
otherwise it'll
( ) stick you

Ms. L: *((In a high thin voice.))* Let it do it by itself.
((Dr. Casaccio explains that he needs to get a culture to make sure of his diagnosis.))

Ms. L: *((Her voice louder and trembling.))* You don't know what they are?

Dr. C: Well I don't—
They're not really
specific for herpes
but they could be
and so this way (will) tell what they are.
Don't move now.
/
(Don't)
touch yourself
you'll spread the herpes.

(May 8, 1987, Lefkowitz transcript, 69–70)

By obeying, she becomes complicit in making herself object, in taking out of the flesh its expressive properties, its signs of itself as animate, incarnate, embodied. Even as her body dys-appears to her, she is induced to assist in its disappearance.

Touch in support of the specific practices of the examination is produced by the gynecologist as instrumental: the hand conducts intelligence to the mind. And it is perceived by the patient as a disturbance to be tolerated, stood still for. The medical body, its animacy and expressiveness muted, its perceptions deflected, is not allowed to move into a rhythm with its interlocutor, to approach pleasure. Touch is neither trespass nor impropriety. The body the gynecologist touches has been rendered object.

So seamless is this rendering that even acknowledging the affinity between gynecological examinations and sexual intercourse may not create an impropriety.

Ms. J: *((During her vaginal examination, Evie Jones winced. As they talk afterward, Dr. Casaccio explains to her that her uterus is out of place.))*

Ms. J: Well what's been
bothering me
is that only in intercourse it hurts.
/

Dr. C: Yeah, well, he does the same thing as I was doing, you know.
He moves the cervix and the uterus around.

(May 1, 1987, Lefkowitz transcript, 7)

THE SPLIT SUBJECT

The parts of the patient's body to which the gynecologist has visual and tactile access are within the enclosure bounded by the gown and drape. The woman, by contrast, has tactile experience of just those parts of her own body from which she is visually excluded. Within the enclosure, her body is thematized, brought to awareness by the touch of the gynecologist. Without the enclosure, her body is effaced. The primary perceptual modality which expands the sense of self out of the body, visuality, is focalized for her there. The perceptual modality which contracts her to her body, touch, is confined to a region to which she does not have visual access. Thus the woman is inveigled into substantiating two realities corporeally. One, a discourse of social subjectivity, mapped onto her upper outer body, engages her visually; the other, a discourse of medical objectivity, mounted on her lower inner body, engages her tactually. She is obliged to construct a split orientation: she is bound tactually to the realm from which she is visually excluded. Pain is occluded in the medical body.

The drape across the patient's lower body also partially encloses the gynecologist in an olfactory realm consisting not only of scents exuded through the patient's skin or her vaginal mucosa but also of the smells of the gynecologist's chemical preparations. Tenting the unclothed lower body is designed not only to confine this chemistry to the realm of medicine but also to keep the woman from perceiving the gynecologist's inspection as an intrusion into her body boundaries or, alternatively, from experiencing her body as intruding on the gynecologist. As Alain Corbin points out, once bodily propriety entailed erasure of bodily odors, "excrement and bodily effluvia were among the ways in which the territory of the 'I' could be violated; they became encroachments" (1986, 163). Wisps of these scents nonetheless waft intermittently out of the enclosure to the woman's nostrils. But the scents are experienced differentially. For the woman, the aromatic, chemical, clinical scents predominate; her own body smells recede. For the gynecologist, the scents of skin, of vaginal secretions, of deposits in the folds of external tissues, float up through the foreground of accustomed chemical odors. Smells, disrespectful of boundaries, send tendrils of one realm into the territory of the other.

Sounds, too, travel unhindered between realms. The evidential boundaries thus effaced can be recuperated by what Marie-Laure Ryan calls "illocutionary boundaries" which "delimit speech acts within a text or a conversation" (1991, 176). Forms of talk can thus be used to reconstitute ontological difference. Talk between the gynecologist and the nurse within the enclosure is earmarked by an "arcane medical terminology which," according to Candace West, "mystified their patients" (1984, 24). The intent of technical jargon is to inscribe objectivity on the body; its effect is to exclude the woman from the discourse. The inaccessibility thus iterated is reiterated by the gynecologist and the nurse by lowering their voices, directing them toward each other, and using what the linguist Malcah Yeager, calls "aside intonations" (1978, pers. comm.).

((Dr. Casaccio, inserting a large swab dipped in vinegar through a space in the speculum speaks to Ms. Rabinowitz.))

Dr. C: O.K., now I'm going to place a little bit of vinegar on you.
O.K., now we can look in.
((To nurse.)) O.K., let me have the ( ).
((Dr.Casaccio uses a long-nosed reverse tweezers to check the patient's cervix.))
O.K., I'm looking inside your cervix.
( )
((To nurse.)) You hold this just like that
right there. *((Hands her the tweezers.))*
Just a little bit forward.
You got it?

Nurse: Mmhm.
Dr. C: O.K., I'm going to take a photograph.
((Dr. Casaccio swings a camera on a long neck around to the vaginal opening and snaps a picture.))
O.K., you can (leave go of it).
A little iodine.
O.K., just
lay this down on the
tip so I can use it again.
Cause I'll be using it.
Uh—
Oh.
No, no, like that.
So that this hangs over ( ).
(Make sure you) don't contaminate (it so I can't) use it.
(Give me) one more ( ).
/
((To patient.)) You all right?
Ms. R: Mmhm.

(April 29, 1987, Lefkowitz transcript, 11–12)

Because these locutions remain quite audible to the patient, they are delivered with respect to her as asides, so that overhearing the doctor and the nurse takes on the aspect of eavesdropping. It is as if the patient were permitted to overhear remarks to which, as overhearings, she is not permitted to respond.

((Dr. Casaccio has just completed a pelvic examination of Jane Wolfe.))
Dr. C: Vagina looks normal.
I don't see—
the only thing I worry about is this
(counter) aspect of the—
the vulva.
((To nurse.)) Let me have some vinegar on a four by four.
Soak it up and I'll just
put that against (her).

Ms. W: ( )
((The nurse puts vinegar on the gauze; the gynecologist puts a rubber glove on his right hand and explains what he is going to do to the patient as if she had not heard the previous interchange.))
Dr. C: I'm going to place a little uh
gauze soaked in vinegar against you.

(April 29, 1987, Lefkowitz transcript, 21)

Although the patient has auditory access to this technical talk, she does not have speaking rights in it. She has ceded entitlement, in Amy Shuman's sense of "understanding communication with respect to ownership of experience—both the experience referred to in the message and the experience of the communication itself" (1986, 18) to the gynecologist.

The patient is invited to participate verbally in the realm of medicine by describing her pain. The gynecologist pitches inquiries about her interior and its history out of the enclosure to the patient. Occasionally he lifts his head up above the rim of the drape or tweaks it down between her knees to catch her eye. In response, she transforms the tactile into the verbal. As Scarry puts it, "If the felt-attributes of pain are (through one means of verbal objectification or another) lifted into the visible world, *and if the referent for these now objectified attributes is understood to be the human body,* then the sentient fact of the person's suffering will become knowable to a second person" (1987, 13). In doing so, however, the person has lifted the pain away from her body. Pain becomes an artifact of discourse to which the gynecologist attends, as it were, in translation. He regards pain not as a sign of the presence of a self but as a clue to the condition of the object. During her vaginal examination, Ruth Quinlan cries out with pain several times; Dr. Casaccio responds by reassuring her that she does not have a cyst on her ovary (May 8, 1987, Lefkowitz transcript, 55–56). The reconstitution of one modality in another, of touch in talk, shifts pain from one realm to the other, from the ordinary realm to the medical one and also transmutes it from an experiential sensation into an analytic remark.

The gynecologist also verbalizes the tactile: he recounts to the patient her own sensations, thereby preempting them.

((As the nurse puts vinegar on a gauze pad, Dr. Cassacio continues his conversation with Ms. Wolfe.))

Dr. C: I'm gonna place a little uh
gauze uh soaked in vinegar against you.

Ms. W: Will it burn?
Or—

Dr. C: Naw, it shouldn't burn, just feel
uncomfortable.
((Dr. Casaccio holds the pad against her vulva.))

Dr. C: O.K., now let your legs fall all the way apart.
Burn?

Ms. W: No.

Dr. C: Just cold.

Ms. W: Yeah.

Dr. C: You painted your toenails, huh?

(April 29, 1987, Lefkowitz transcript, 21)

With this last remark, the gynecologist shifts the woman's attention one more remove from her body.

The gynecologist's verbal forewarnings of upcoming tactile sensations also appropriate the tactile to the verbal. Just before the gynecologist examines JoAnn Becker's vagina, he says, "O.K., now I'm going to—," touches her vulva with one hand, and inserts the speculum with the other (May 6, 1987, Young transcript). I assume his touch of the vulva is intended to take a bead on the aperture in order to decide the angle of entry for the speculum but it is experienced not as a forewarning but as a prognostication which controls what then transpires.[5] Verbal descriptions of gestures as preludes to the gestures themselves also subsume touch under the modality of talk.

By far the most common arrangement is for the gynecologist to maintain with the patient a state of talk about something other than the examination throughout its course. The gynecologist constructs the realm of medicine visually and tactually and at the same time verbally constructs another reality conjointly with the patient. The patient participates verbally in the construction of this joint reality while she is tactually tugged into the medical realm from which she is visually excluded.

((Throughout her internal examination, during which the gynecologist is palpating the interior wall of her vagina with his fingertips, Meredith Ashbury talks about the broadcasting business.))

Dr. C: What's CBS up to now?

Ms. A: NBC.

Dr. C: Or NBC.

. . .

Ms. A: Making money. *((On this remark, the gynecologist inserts his hand into her vagina.))*

(May 8, 1987, Lefkowitz transcript, 10)

The alternate reality may in fact be medical, but not the medical reality at hand, thus preserving a thematic relevance along with a practical exclusion.

((Throughout her pelvic examination, bracketed by the gynecologist's paired instructions, "Come on down," and "O.K., scoot back," Dr. Casaccio and Erika Christianson talk about her mastectomy.))

Dr. C: Come on down.

((The patient slides down to the foot of the table. Over the course of ensuing conversation, the gynecologist puts on gloves, inserts the speculum, and does a Pap smear.))

/

What's it been now, two years?

Ms. C: No.

Dr. C: One.

[[[ ]

Ms. C: It was a year at Christmas.

Yes=

Dr. C: O.K.=

Ms. C: Mmhm.

/

Dr. C: Well that's really a— a shock to
to go through that—

Ms. C: I feel like I've
made a
good adjustment

Dr. C: Oh yeah well now it's getting=
Ms. C: you know=
Dr. C: further away
and I think you feel a little bit more=
Ms. C: Mmm.=
Dr. C: secure.
But you know when it first happens
you know
your whole security is gone.
Ms. C: Well I (had) to
have the uh pulmonary embolism at the same time.
It was (   )
[[[   ]
Dr. C: It wasn't easy. *((The gynecologist has removed the speculum and, with this remark, inserts his hand into the patient's vagina to do the internal examination.))*
Ms. C: It took me a good six months.
Dr. C: Even the family probably had a hard time.
Ms. C: We had a rough year *((She chuckles.))*
/ *((During this pause, Dr. Casaccio inserts his finger in Ms. Christianson's anus, removes a trace of fecal matter, and wipes it on a slide.))*
Dr. C: *((To nurse.))* Let me have a stool.
/ *((The nurse holds out the slide.))*
O.K., scoot back. *((He tucks his hands under her buttocks to guide her to sit up.))*

(May 8, 1987, Lefkowitz transcript, 23)

The patient's split orientation creates a break between pain and its surrogates, secreted in the body, and either the wince or its surrogates, erupting on the body, or the whimper and its surrogates, erupting in the voice. In the realm of medicine, pain is mentionable but not expressible.

The woman may contrive to remain oriented to her tactual engagement in her own body by attending verbally to its examination. But this attention is, at best, indirect and, at worst, complicit in her own objectification. Its indirection lies in the transformation of the tactile into the verbal, which eludes the hold the body has on her mind, its visceral insistence. She is obliged to set aside tactual experience, her sense of herself as embodied, and address herself as an object. Scarry regards this, not as the beginning of inauthenticity but as the beginning of invention. Articulating pain, even as it dematerializes the body, materializes for the woman in pain another reality. Scarry writes, "to be present when a person moves up out of that pre-language and projects the facts of sentience into speech is almost to have been permitted to be present at the birth of language" (1987, 6). But the act of articulating moves the person beyond her sentient experience: "*expressing* physical pain eventually opens into the wider frame of *invention*" (1987, 22). In speaking, the woman retrieves the contents of the world dismantled by pain, including her own body as a sentient presence in that world. We are witness to what Scarry calls the making of the world (1987, 23).

((As he is palpating her abdomen, Dr. Casaccio asks Evie Jones about a pain she mentioned.))

Dr. C: Where's this pain?

Ms. J: Right here but my whole abdomen's been cramping.

This remark lifts the pain away from her body by articulating it and at the same time retrieves her body as a sentient presence in the realm of medicine. Subsequent talk holds to the contours of patient's and gynecologist's joint experience of her body.

Dr. C: Something there.
I can feel the pulse in your artery but
that may—
/////////////
I can't feel anything.
///
I think it's just close to your backbone *((hiccup in the tape))* feel your—
feel your heartbeat.
/
This is the artery.

Ms. J: Well something's there but once in a while something's—

[[[ ]

Dr. C: Let me check you (below) see how you feel.
Ms. J: something's hard right there occasionally.
Dr. C: Really?
Ms. J: But I don't know what—
It goes away.
/
((Dr. Casaccio moves round to sit on his stool at the foot of the examination table and gloves himself. His nurse turns on the spotlight and aims it at the patient's vaginal opening.))
It's not like it stays there and it's anything to worry about.
/
Least I don't think so.
Dr. C: *((As he touches her vulva))* Still riding your horse?

(May 1, 1987, Lefkowitz transcript, 2)

This remark, uttered at the moment the gynecologist touches her body, splits the woman's attention off from her embodiment on the occasion of the examination and redirects it to her embodiment elsewhere.

The gesture of representation, of representing the body's immediate experience verbally, positions the woman to make other representations, representations of her body's remote experience, and to make these as part of her sense of herself on the occasion of her examination. She is poised to tell stories, to recount a world in which she is sovereign even as she is perceptually bound to a world in which she is, in Julia Kristeva's term, abject. Her enclosure in the realm of medicine cannot wholly protect her from the examination's effacement of the boundary between her continent outer surface and her overflowing interior. She is exposed to the uncanniness of her viscera, preventing her under the circumstances of the examination from perceiving herself as intact, as her proper self, by which Kristeva means both her own self and her clean self.

> The body's inside, in that case, shows up in order to compensate for the collapse of the border between inside and outside. It is as if the skin, a fragile container, no longer guaranteed the integrity of one's "own and clean self" but, scraped or transparent, invisible or taut, gave way before the dejection of its contents. (Kristeva 1982, 53)

Her abjection, in an insight Kristeva traces to Georges Bataille (Kristeva 1982, 64), inheres in her inability to exclude the encroachment, which the gynecologist effects, of the abject viscera.

Shifting her locus of consciousness from medicine to narrative, as the realm in which to spin out a presentation of self, can disrupt the dominance of medical discourse over the voice of the lifeworld. Physicians characteristically treat storytelling as an interruption of, distraction from, or incursion into the realm of medicine. But the shift can also release the woman from her incarnation as a patient. Embodiment elsewhere, in narrative, permits disembodiment here, on the occasion of the examination. And this, the gynecologist may abet. The condensation of a narrative self assists the attenuation of her tactile embodiment. Disembodiment supports the general medical project of objectification that is requisite for the specific gynecological attempt to ensure propriety. Tactility especially, as it thrusts her own body upon the patient's notice, resists objectification. Dr. Casaccio, therefore, invokes this ontological shift in Evie Jones at just the moment he touches her body. As he touches her breast, the doctor asks,

Dr. C: How's your legal
battles?

(May 1, 1987, Lefkowitz transcript, 1)

As he touches her vulva, another question,

Dr. C: Still riding your horse?

(May 1, 1987, Lefkowitz transcript, 2)

And as he inserts his hand into her vagina,

Dr. C: When were you in South Africa?

Thus the gynecologist occasions the creative act by which the woman can reconstitute herself. The woman becomes a split subject: the realm in which she is sentient is closed off from the realm in which she is articulate. Yet, out of the pain that binds her to her body, she can fabricate the imaginative act that lifts her out of her body and into her reconstituted self.

Despite the gynecologist's invitation to do this, the women Dr. Cassacio so addresses never seem to produce a story. Their usual disposition is to politely withdraw themselves from their bodies, to erase their social personae for the nonce and accede to the conventions of objectification constructed for the examination: visual exclusion, verbal silence, and tactile suppression; coupling this with alertness to cues to enter into medical discourse, not as an embodied self but as a fellow observer, if a uniquely positioned and uniquely knowledgeable one, of quirks of the flesh. Signs of this angle of entry, as it were from outside her own body, appear in the desuetude of such deeply embodied apprehensions as pain and pleasure, the tactile perceptions which attach body to world. Patients instead thoughtfully set aside their preoccupations with the lifeworld and give themselves over, lend themselves, to the realm of medicine. This particular gynecologist attempts to retrieve their lifeworld for them, not just to give them another footing on the occasion of their examination, but to provide distraction from its attendant pains. Dr. Casaccio aims, not for sightlessness, senselessness, and silence, but beyond these to an alternate reality. Why do women never invoke the narrative domain he invites?

NARRATIVES OF INDETERMINACY

In fact, storytelling is exceedingly rare on any medical occasion. Technically speaking, patients fail to produce at least two necessarily sequenced clauses of which the second is consequential on the first. William Labov and Joshua Waletzky's work elaborates this minimal criterion as the core gesture of narrative (1967; Labov 1972; see Young 1986, 28–29). Instead, patients produce what Erving Goffman calls "replays" (1974, 504) that do not achieve the status of fully-fledged narratives, stories that do not come to an end, the sorts of stories Labov calls "so what" stories (1972, 366). In so what stories, one thing happens after another; in narratives proper, one thing follows on from another. The difference is between events that are merely incidentally sequentially ordered and events that are consequentially related (see Young 1986, 206–209).

Consequentiality is what constitutes one of a series of narrative clauses the end of the story. In effect, stories are constructed backward: tellers start with the end and recoup the sequence of events that leads up to the end as the story. The first of these necessarily sequenced clauses becomes the beginning. In consequence, "beginnings do not so much imply ends as ends entail beginnings" (Young 1986, 29). Stories achieve a kind of redundancy that knits together a sequence of events in a relation hearers take to be causal. Paul Ricoeur writes, "By reading the end in the beginning and the beginning in the end, we learn also to read time itself backward, as the recapitulating of the initial conditions of a course of action in its terminal consequences" (1980, 180). By virtue of this construction, the events in stories appear to come to completion. Consequentiality imparts a direction to events, providing them with a temporal axis and a causal logic. The directedness of events, which takes the appearance of temporal unfolding in tellings, is in fact an effect of the atemporal enfolding of ends in beginnings. But in replays, there are no beginnings and ends, only ongoing events. So events do not appear to take a direction. As a consequence, hearers cannot locate the temporal horizons of the taleworld. Maurice Natanson writes, "Taking the world as an outer limit, its horizon is the condition for bounding what can be experienced" (1962, 89), the physical or metaphysical limits of its universe.

The absence, lack, or failure of temporal closure in the taleworld opens out into another absence. Replays fail to create an enclosure within the realm of conversation for discourse of another ontological status, in this instance, narrative discourse. In becoming patients, persons set aside some of their claims as social interactants, among them the right to take an extended turn at talk in order to tell a story. Instead, patients cede physicians exclusive rights to initiate conversational sequences on the assumption that the resultant speech events will suit the requisites of medical practice. This arrangement creates an inequality between physician and patient under the aegis of ensuring the efficiency of the examination.

A turn at talk is protected, sealed off, in principle, from interruptions by prospective other speakers. Within the turn, narrative discourse forms an enclave of a different ontological status within the realm of conversation, a storyrealm in which the storytelling transpires. Despite the turn and the realm-shift within it, conversational anecdotes tend to remain pervious to remarks by other conversationalists that may or may not count as interruptions. Replays consist primarily of narrative clauses which are sequentially ordered but not consequentially related. Each event recounted might or might not be followed by a next event. Since a next narrative clause does not complete the replay, that is, make the replay into a story, so the replay is not incomplete without it. If all the narrative clauses that make up a story are entailed in a single turn at talk (and this can be so despite interpolations by other speakers), by contrast, each narrative clause of a replay constitutes by itself a single turn at talk which is potentially complete. Replays remain open to remarks by other speakers which are never interruptions. Thus the boundaries of both realms, the realm of the events the story is about, or taleworld, and the realm of narrative discourse within conversation, or storyrealm, remain pervious. Neither the event nor the story is sharply defined off as a separate reality.

In their conversation, Dr. Casaccio invites Evie Jones three times to embark on stories, each time at the moment that engendering a narrative self might lift her away from an uncomfortable body, the story of her legal battles, the story of her horseback riding, and the story of her adventures in South Africa. The first of these transpires right after the opening sequence of the physical examination. As is typical of medical examinations, the remark that might have been produced by the gynecologist as a greeting ritual, "What's going on?" is treated by the patient as a medical inquiry, making it the pivot for shifting from the realm of the ordinary to the realm of medicine. The gynecologist shows a commensurate attention to the joint construction of the realm of medicine by chiming in to the patient's response to his own question with "but they hurt like mad." Chime-ins are one of a class of phenomena Harvey Sacks refers to as "collaborative productions" or sometimes "joint productions" (Sacks 1992, I, 321; II, 57). Like all such phenomena, they display the shared understandings which construct a joint reality.

Dr. C: Evie.
What's going on.
[[[ ]
Ms. J: Hi again. *((To the researcher.))* (Heh.)
Um
Same old thing. *((To the gynecologist.))*
Dr. C: Have your periods?
Ms. J: Yeah?
but
Dr. C: =they hurt like mad.
Ms. J: Right.
This side right here hurts again.

Dr. C: All right, let me take a look atcha.
///
((During this pause, Dr. Casaccio has Ms. Jones lie down, lifts her right arm over her head to rest on the cot, and moves her gown aside to begin the breast examination. Over the course of their subsequent talk, he conducts the examination, moving from her right to her left breast, then to her abdomen. During this time the gynecologist looks intermittently at the patient's breast and belly; the woman turns her gaze to the ceiling. Just as he touches her breast for the first time, Dr. Casaccio makes a remark.))

Dr. C: How's your legal
battles.

Ms. J: Um,
I don't know— we don't know if we'll have a house
anymore or not.

[[[ ]

Dr. C: Oh yeah?
That's terrible.

Ms. J: Tell me about it.
But

Dr. C: Lawyer came and assessed everything in your property?

Ms. J: The lawyer himself said he did— Ron was ready— Ron was
he was out there ready to—
he was pointing his— Justin said, second time?
((Justin is the lawyer.))
He goes,
"Oh I've had a death in my family I have to leave" I said that's what you said last time.

Dr. C: (Hehehe) uh (he) ah ah.

Ms. J: He says the same thing every time when he gets in trouble? and he has to leave quick?
"Had a death in my family I have to go."

All: *((All laugh.))*

(May 1, 1987, Lefkowitz transcript, 1)

Dr. Cassacio's remark, "How's your legal battles?" licenses Evie Jones to take an extended turn at talk in order to tell a story. Jones's response, "I don't know— we don't know if we'll have a house anymore or not," is an abstract that could serve either as an answer to his question or as a preface to the story. Dr. Casaccio's attention marker, "Oh yeah?" inflected as a question, serves as a comment that might also invite a continuation. The paired evaluative remarks, his "That's terrible" and her "Tell me about it," both offer appreciations of the import of her abstract and so could constitute closure on this sequence (H. Sacks 1992, II, 125; Young 1986, 88). The gynecologist's proposed abstract, though, "Lawyer came and assessed everything in your property?" initiates narration in the form of a report of what the lawyer said to Evie's man, Ron. The story consists of two narrative clauses separated by temporal juncture, what Ron did and what the lawyer said, the minimal narrative. Speaking constitutes an act in the taleworld but what is spoken does not. "Had a death in my family . . ." invokes a world laminated by the verb "goes" (Goffman 1974, 505) within the world of the tale, another realm of events without consequences for this taleworld. The shift of reported speech from past to present, evidenced in the shift of the laminator verb from "said" to "goes" is, however, typical of the way narrators embody themselves in the taleworld as a present reality at the climax of a story (Goffman 1974, 508). Evie Jones's rejoinder, "That's what you said the last time . . ." transforms what promised to be the story of her legal battles into a story about one-upmanship. Since the consequences of the lawyer's activities are not recounted here, the story becomes an episode in the longer narrative of the legal battles. The lawyer's departure appears to delay resolution of the longer story to some subsequent episode.

Replays like these, following an argument of Mark Workman's about narratives within what he calls paradigms of indeterminacy (1993), I should like to call "narratives of indeterminacy." Such open-ended narratives are not to be regarded as flawed, either failed deliveries or failed receptions, but as intentionally incomplete. Their incompleteness is aimed in two directions. One, a strategic direction, is the organization of the narrative to span a piece of interaction, at the conclusion of which it can be abandoned or cut off without counting as having been interrupted. And two, a constructionist direction, by refusing consequentiality as an organizing principle, narratives of indeterminacy dismantle temporal unfolding and its underlying assumption of causality. Workman and others (Nicolaisen 1991, 9) have speculated that the recursive character of narratives of indeterminacy puts forward spatiality rather than temporality as the organizing principle of narrative. On the one hand the realm of medicine constrains the production of narrative discourse and, on the other, the narratives produced in turn construe indeterminate realities.

The gynecologist's touch of the breast, the abdomen, or the vulva serves gesturally as the opening frame of each of the three phases of the physical examination. With each gesture, the gynecologist enters into an examination of the body. In doing so, he changes what Charles Goodwin refers to as the "engagement framework" of the interaction (1981, 10). Up to the first gesture, the gynecologist and the woman have displayed mutual orientation to their engagement. Once the physical phase of the examination begins, participants "move from a state of talk to a state of disengagement" in order to permit the exigencies of the examination to take precedence. Goodwin points out, "After disengagement has been entered, talk is still possible, but this talk has both a different sequential organization at its boundaries and a different structure of coparticipation in its course than talk produced during full engagement." Nonetheless, he notes that "although during disengagement the participants are explicitly displaying nonorientation toward each other, each is in fact paying close attention to what the other is doing" (Goodwin, 1981, 10).

In putting questions to the patient just as he touches her body, the gynecologist is opening the state of disengagement to the sort of loose talk Goodwin mentions. When the questions are about her lifeworld, the gynecologist proposes himself as what Goodwin calls an "unknowing recipient" of some proposed piece of talk (1981, 10). The woman is then expected to produce an utterance that imparts to him new information. This appears to be designed to ensure that the woman is taken up with producing talk while the gynecologist attends to her body. To sustain her production, the gynecologist need only evidence hearership. Goodwin notes that "one way in which a nonspeaking party can indicate whether he is acting as a hearer is by gazing at the speaker" (1981, 9). But mutual gazing has been dismantled in the disengagement framework. "Hearership can of course be demonstrated in other ways" (Goodwin 1981, 9), such as attentive vocalizations, tag questions, comments, or evaluations.

When these are not forthcoming, the patient can suppose that the gynecologist has invoked the disengagement frame and turned his attention to the examination. Within this framework, the patient can always be recalled, tactually or verbally, to her own examination. Hence, embarking on a story risks interruption.

One solution to the problem of sustaining talk within the disengagement framework is to direct talk to the examination, the proposed primary focus of interaction. And indeed the gynecologist occasionally addresses the questions with which he opens either phase of the physical examination to the examination itself. Questions like, "Are you having any problems?" as he touches the patient's breast (May 8, 1987, Lefkowitz transcript, 31) or "Did you have any problems after the operation?" as he touches the site of the operation, the woman's vagina (May 8, 1987, Lefkowitz transcript, 43), license the patient to participate in medical discourse. Besides obliging the patient to translate tactile experiences into verbal modalities, the question also obliges the woman to recount what has happened to her body historically rather than attuning to what is happening to it currently. Her account is topically connected to her bodily experience but temporally disconnected. Permitting the patient to participate in medical discourse makes the gynecologist's gate-keeping job more difficult. The elaborate constructs designed to exclude her from that realm are set aside. In a rare instance in which the gynecologist provides the patient a minute-to-minute account of what he is doing in the examination, such as Dr. Casaccio's hunt for the source of the pain in Evie Jones's belly, he does tune his talk to her experience but keeps it under his control. She does not talk. If, in these excursions into ordinary discourse, the gynecologist proposes himself as the unknowing recipient, in medical discourse, the patient is generally proposed as the unknowing recipient.

Some women embark on accounts of other medical experiences that apparently maintain topical relevance to the realm of medicine without intruding on their own examinations. This solution has the curious property of making the gynecologist the unknowing recipient of information about his own domain. In one instance in which a woman was recounting to Dr. Casaccio the story of the premature child of a friend of hers, this reversal apparently became so uncomfortable that, in spite of the gynecologist's attentive comments, in the middle of her story, she switched the gynecologist to knowing recipient.

Woman: But I— I— I hope—
Where did—
Doctor Casaccio
do you think the one that's in the hospital will be O.K.?
Will it ever be normal?

(May 1, 1987, Lefkowitz transcript, 21)

Persons proposed as knowing recipients participate in talk by telling the speaker what they know; persons proposed as unknowing recipients participate in talk by indicating attention to the speaker's talk. Since signs of hearership can be subtle or absent in this latter instance, the patient transforms the gynecologist from unknowing to knowing recipient in order to ensure his active participation.

Most of the questions Dr. Casaccio times to coincide with the moment he first touches the patient's breast, abdomen, or vulva, or when he first introduces his fingers into a patient's body, direct her attention away from the occasion of the examination and toward the lifeworld in which she can be presumed to exercise her sovereignty. Some of these questions are designed to address what the gynecologist takes to be the woman's circumstances. These range from generic questions like "Anything happening?" to such circumstantial questions as "Getting all ready for the summer?," "Where do you go to school now?," or "How's the family doing?," (all May 8, 1987, transcript, 28, 47, 29, 22). Other questions rely on the gynecologist's knowledge of the woman's lifeworld. Questions like "How's your legal battles?," "Go flying with your father at all?" (May 1, 1987, transcript, 1, 30), or "What's CBS up to now?" (May 8, 1987, transcript, 10) not only invite stories about the woman's lifeworld but also attest to the gynecologist's attention to such stories on prior occasions. The invitation keeps the patient from intervening in her own examination at just the moment her body dys-appears in her awareness, and at the same time offers her another footing for presenting a self.

The gynecologist's conversational moves toward storytelling rarely eventuate in patients' stories. Instead, many patients take these invitations as questions to which they should make short answers, thus keeping the floor clear for proper medical inquiries. Those who do appear to embark on stories are often recalled to their corporeal circumstances either by a medical question or by a pain in their bodies. Because the beginnings of stories aim for their ends, the introduction of consequential relationships between narrative clauses lays the story open to interruption if the patient is recalled to the realm of medicine before its end. For this reason, patients characteristically produce narrativelike speech events or replays in the form of episodic accounts whose rambling structure could be broken off at any point without counting as interruptions.

This first episode of Evie Jones's account of her legal battles closes with a replay whose last phase is the speech act which constitutes a leave-taking, "Justin said, second time?, he goes, 'Oh I've had a death in my family I have to leave.' " Without pausing for breath she appends her response, "I said, 'That's what you said last time.' " The event that precipitates this sudden departure has been elided in the telling but the replay offers a clue to what it was. "Ron was ready— Ron was— he was out there ready to— he was pointing his—." What do you suppose he was pointing? My guess is, a gun. If so, the patient's elision of that fact from her telling may be intended to preserve her presentation of herself as the lawyer's victim rather than he hers. The lawyer's response, skedaddling, does not end the story of the legal battles but rather delays its resolution. With the gun elided, the lawyer's departure is not the conclusion of a course of action but the extension of that course into some future account. Indeed, the lawyer reappears in a subsequent episode of Evie Jones's life story, one that likewise detours resolution and one from which she is recalled by a visceral twinge.

((Dr. Casaccio comments on the death-in-the-family line.))

Dr. C: That's a polite way of saying I do(hoho)n't want to stick around and face the heat?

Ms. J: Yeah.

Dr. C: =Yeah. ((Whispers))

/

Ms. J: Well I was pretty upset when
he came out and said um
that he'd been on my property?

Dr. C: Yeah that was worth X amount of dollars.

[[[ ]

Ms. J: Kind of

/

Dr. C: I don't know.

/

((Dr. Casaccio is now palpating Evie Jones's abdomen. She directs her gaze straight ahead of her; his gaze is angled up under his eyebrows, abstracted. Neither looks at her body nor at the other.))

Ms. J: He's definitely different as a lawyer than he is as a person= Ouch that's the pain.

(May 1, 1987, Lefkowitz transcript, 2)

The indeterminate narrative becomes an episode in a presumably ongoing saga on which the teller reports periodically. Although events within each episode are sequentially ordered, the episodes are not necessarily sequentially ordered with respect to one another. It is rather as if the taleworld provides a spatial anchorage for various narrative presentations, over the course of an occasion or a number of occasions, of a self who is unfolding. The storyteller produces fragments for what might be called a history of the *postmodern* body, to adapt the title from *Zone* (Feher 1989), an incomplete, discontinuous, metamorphic, ramifying body.

Since narratives of indeterminacy never come to an end, breaking off in the course of them cannot be construed as an interruption. These narratives are not closural but improvised flexibly to span the rhythm of the examination as it transpires in the realm of medicine. Such narratives are open-ended, episodic, ongoing, and recursive or temporally indeterminate. Instead of a temporal unfolding, the narratives take up a spatial anchorage. So women solve the problem of the interruptibility of their narratives within the disengagement framework by traversing the space of a narrative realm, a taleworld, without fully inhabiting it, without being enclosed in it. Instead, they keep the taleworld open for quick visits, brief sorties which give them a foothold in another world.

BREAKING THE DISCURSIVE BODY OUT OF POSTMODERNISM

In gynecological examinations, the body is, as it were, lifted out onto different planes of reality, reconstituted in different universes of discourse. But these are not instances of the postmodern dematerialization of the body into its signs. N. Katherine Hayles remarks, "One belief from the present likely to stupefy future generations is the postmodern orthodoxy that the body is primarily, if not entirely, a linguistic and discursive construction . . . discourse theory, . . . information theory, . . . information technologies . . . collaborate in creating the dematerialization of embodiment that is one of the characteristic features of postmodern ideology" (1993, 147). Dismantling the body into its discourses does not dematerialize the body but rather has its footing in embodiment. Hayles writes,

> In contrast to the body, embodiment is contextual, enwebbed within the specifics of place, time, physiology and culture that together comprise enactment. Embodiment never coincides exactly with 'the body,' however that normalized concept is understood. Whereas the body is an idealized form that gestures toward a Platonic reality, embodiment is the specific instantiation generated from the noise of difference. Relative to the body, embodiment is other and elsewhere, at once excessive and deficient in its infinite variations, particularities, and abnormalities. (1993, 154–155)

In these gynecological examinations, one discursive body is substantiated visually, another is substantiated tactually, and the third is substantiated narratively. Personhood is constituted differently in each perceptual modality. And in them all, the body, as it is thought or felt, turns out to have a foothold in embodiment.

THREE

• • •

DECIPHERING SIGNS • SURGERY

It would be better to speak of a certain "bodiliness" than of "the body." It is the instance of a suturing of discourse and desire to the organism . . .

—*Francis Barker*

What it is to be a person is a cultural notion, one of the epiphenomena of ideology. The term "ideology" captures in contemporary social scientific usage aspects of what used to be called "metaphysics": the hold I have on what I take to be reality. Louis Althusser, magisterially cleaving the ideal from the material, argues, specifically, "Ideology represents the imaginary relationship of individuals to their real conditions of existence" (1971, 162). Ideologies, one might say, invent their own subjects. I know what it is to be a person in my own realm, to find myself at a nexus of tensions and pressures in cultural space and likewise to find myself constrained to be there. My subjectivity, my sense of myself, is at root ideological and in this sense imaginary, fabricated.

MAKING PERSONS

So personhood is a notion with consequences. The way I think myself and see others is tangled up in the social fabric. Ideology is a sorcery to which I am apprenticed. I conjure up myself, others, objects, worlds enough and time. I conjure us up bodily. We substantiate, in our own persons, the ideology we invent. As Althusser puts it, "All ideology hails or interpellates concrete individuals as concrete subjects" (1971, 173, emphasis omitted). The virtue of this "hailing" or "interpellation" is to materialize, to embody, a turn of mind, an act of the imagination, a gesture of invention.[1] Because the body as a locus of self is imaginary, the embodied person must be constantly reconstituted, reinvented, *intended,* to take up the phenomenological term. I am complicit in the invention of my own subjectivity. The inscription of ideology on materiality (Althusser 1971, 165), the self on the body, gives this invention its footing in the real.

Cultural practices, then, inscribe the body into the discourse of subjectivity as the locus of the self. Thus ideology appears to be conjured up out of materiality. The body is put forward as the source of the self, which comes to be regarded as a sort of bodily effluvium, either a preexisting essence perfused through the substance of the body or an ether effused out of it. The bodily self is a socially constituted performance, sustained as a reality, of the sort that Erving Goffman calls a "presentation of self" (1959).

> A correctly staged and performed scene leads the audience to *impute* a self to a performed character, but this imputation—this self—is a *product* of a scene that comes off, and is not a *cause* of it. The self, then, as a performed character, is not an organic thing that has a specific location whose fundamental fate is to be born, to mature, and to die; it is a dramatic effect arising diffusely from a scene that is presented. (1959, 252)

Social life is conducted as a series of appearances in which I evidence or attest to the sort of self to which I lay claim and which I thereby fabricate. Having or being a self is not a condition of life; it is a social accomplishment. The person "and his body merely provide the peg on which something of collaborative manufacture will be hung for a time. And the means for producing and maintaining selves do not reside inside the peg" (Goffman 1959, 253). Subjectivity inheres in the social fabric and is only incidentally localized at the nexus of the body.

The attempt to bind the self to the body conceals the Lacanian split, the discovery of the body as object and as Other, as what Jacques Lacan calls the "alienating destination" of the *I.* A moment in childhood attests at once to the solidity of the body and to the constitution of subjectivity elsewhere. I catch sight of myself in a mirror and am thereafter haunted by my image in its various "modalities of materiality" (Althusser 1971, 169) as statue, as phantom, as automaton (Lacan 1977, 2–3). This image of my body, as opaque and elsewhere, as object and Other, "by these two aspects of its appearance, symbolizes the mental permanence of the *I,* at the same time as it prefigures its alienating destination" (Lacan 1977, 2–3). Maurice Merleau-Ponty writes:

> By means of the image in the mirror he becomes capable of being a spectator of himself. Through the acquisition of the specular image the child notices that he is *visible,* for himself and for others. The passage from the introceptive *me* to the visual *me,* from the introceptive *me* to the 'specular I' (as Lacan still says), is the passage from one form or state of personality to another. . . .
>
> To use Dr. Lacan's terms, I am 'captured, caught up' by my spatial image. Thereupon I leave the reality of my lived *me* in order to refer myself constantly to the ideal, fictitious, or imaginary *me,* of which the specular image is the first outline. (1964, 136)

Lacan's "mirror stage" dislocates my sense of my inherence in my own body. I come upon a trace of my subjectivity being constituted elsewhere.

That nexus, that locus, that body, is discursively shaped. It is, as Pierre Bourdieu puts it, a *"socially informed body"* (1989, 110). Its accoutrements, its skills, its gestures and postures, even its dispositions, are composed within a "*universe of practice* (rather than a universe of discourse)" (1989, 110). The body develops a hexis, from the Greek word for habit.

> Body *hexis* speaks directly to the motor function, in the form of a pattern of postures that is both individual and systematic, because linked to a whole system of techniques involving the body and tools, and charged with a host of social meanings and values: in all societies, children are preternaturally attentive to the gestures and postures which, in their eyes, express everything that goes to make an accomplished adult—a way of walking, a tilt of the head, facial expressions, ways of sitting and of using implements, always associated with a tone of voice, a style of speech, and (how could it be otherwise?) a certain subjective experience. (Bourdieu 1989, 87)

Thus the body becomes my body. My subjectivity materializes in a universe of practice. I carry the inscription of culture on my body as corporeal memory.

> If all societies set such store on the seemingly most insignificant details of *dress, bearing,* physical and verbal *manners,* the reason is that, treating the body as a memory, we entrust to it in abbreviated and practical, i.e. mnemonic, form the fundamental principles of the arbitrary content of culture. The principles em-bodied in this way are placed beyond the grasp of consciousness, and hence cannot be touched by voluntary, deliberate transformation, cannot even be made explicit; nothing seems more ineffable, more incommunicable, more inimitable, and therefore, more precious, than the values given body, *made* body by the transubstantiation achieved by the hidden persuasion of an implicit pedagogy, capable of instilling a whole cosmology, an ethic, a metaphysic, a political philosophy, through injunctions as insignificant as "stand up straight" or "don't hold your knife in your left hand" . . . The whole trick of pedagogic reason lies precisely in the way it extorts the essential while seeming to demand the insignificant. (Bourdieu 1989, 94–95)

Body becomes "habitus," a locus for the "*dialectic of the internalization of externality and the externalization of internality,* or more simply, of incorporation and objectification" (Bourdieu 1989, 72). Culture is corporealized; the body is enculturated: "the habitus could be considered as a subjective but not individual system of internalized structures, schemes of perception, conception, and action" (1989, 86), which Bourdieu calls *"dispositions"* (1989, 72). So investing ideology in the body, Francis Barker writes, secures "the deep inner control, which has been called interpellation, that is so vital to the bourgeois illusion of freedom in its civil and its personal guises" (1984, 59).

It is in consequence of this appropriation of materiality by ideology that, as Althusser argues, subjectivity can be represented in the humanist tradition as individuality (1971, 138). The myth of individuality rides on the substance of idiosyncrasy, as if physical particularity generated conceptual uniqueness. In reality, the reverse is the case: I take up physical particularities as marks, signs, inscriptions of a self in order to attest to a uniqueness that is not there. Individuating marks are so described on the surface of the body that personhood appears to thin out over the broad expanses of the body and to condense in its puckers, whorls, slits, nipples, invaginations, and outpouchings. The deciphering gaze skips and scrutinizes, swept over planes and mounds, hollows and hummocks, to stick on the dense, folded, concentrated structures, especially where they cut into the body. Richard Selzer, surgeon and essayist, celebrates just this tinge of subjectivity in surface inscriptions.

> I sing of skin, layered fine as baklava, whose colors shame the dawn, at once the scabbard upon which is writ our only signature, and the instrument by which we are thrilled, protected, and kept constant in our natural place. Here is each man bagged and trussed in perfect amiability. See how it upholsters the bone and muscle underneath, now accenting the point of an elbow, now rolling over the pectorals to hollow the grotto of an armpit. Nippled and umbilicated, and perforated by the most diverse and marvelous openings, each with its singular rim and curtain. Thus the carven helix of the ear, the rigid nostrils, the puckered continence of the anus, the moist and sensitive lips of mouth and vagina.
>
> What is it, then, this seamless body-stocking, some two yards square, this our casing, our facade, that flushes, pales, perspires, glistens, glows, furrows, tingles, crawls, itches, pleasures, and pains us all our days, at once keeper of the organs within, and sensitive probe, adventurer into the world outside? (1976, 105)

Skin is reified as the "cut" between interior and exterior, self and Other, subject and object, signifier and signified (Lacan 1977, 299). But its status remains ontologically ambiguous. The "specular image," within which the Lacanian experience of subjectivity as turbulence (1977, 2–3) is entrapped, is also the "channel taken by the transfusion of the body's libido towards the object" (1977, 319). Susan Stewart writes that this reveals "the paradoxical status of the body as both mode and object of knowing, and of the self constituted outside its physical being by its image" (1984, 131). Its objecthood pinions the body as signified, as lodgment of otherness, of the Other, of the self as Other. Yet the anchorage of perception in the body reconstitutes subjectivity here. My body recurs to me as source of signification, as signifier. The body's "return to the inanimate" (Lacan 1977, 301) is confounded by this recuperation. Subjectivity is represented by turns as interiority and exteriority: the self is sacked in skin or stamped on skin. So there, on the skin, a discourse of interiors is opposed to a discourse of surfaces.

Cuts into the body perforate this surface. They rupture the continence of the skin as container of subjectivity, they blur interior and exterior, they evert the lining, not of the body but of the self. Cuts, Lacan argues, "have no specular image, or, in other words, alterity. It is what enables them to be the 'stuff,' or rather the lining . . . of the very subject that one takes to be the subject of consciousness" (1977, 314). They are sites for the emergence of subjectivity onto the surface of the body, an exteriorization of interiority. Here are the openings of the subject to the Other. Cuts are marks of desire. According to Lacan, desires are associated with organic systems, invisible in the body (1977, 314), except for the "cut [*coupure*] expressed in the anatomical mark [*trait*] of a margin or border—lips, 'the enclosure of the teeth,' the rim of the anus, the tip of the penis, the vagina, the slit formed by the eyelids, even the horn-shaped aperture of the ear" (1977, 314). Ambiguously directed toward both object of desire and source of desire, cuts become what Lacan calls "erogenous zones" (1977, 314).

Cuts also proliferate the surface of the body. Its sheer materiality is adumbrated. It is a body of replication, multiplication, exaggeration, distortion, the baroque production of involutions and obtrusions that are imitations, transformations, deformations of one another. It is Mikhail Bakhtin's grotesque body.

> The grotesque body . . . is a body in the act of becoming. It is never finished, never completed; it is continually built, created, and builds and creates another body. Moreover, the body swallows the world and is itself swallowed by the world . . . This is why the essential role belongs to those parts of the grotesque body in which it outgrows its own self, transgresses its own body, in which it conceives a new, second body: the bowels and the phallus. These two areas play the leading role in the grotesque image, and it is precisely for this reason that they are predominantly subject to positive exaggeration, to hyperbolization . . . All these convexities and orifices have a common characteristic; it is within them that the confines between bodies and between the body and the world are overcome: there is an interchange and an interorientation. This is why the main events in the life of the grotesque body, the acts of bodily drama, take place in this sphere. Eating, drinking, defecation and other elimination . . . as well as copulation, pregnancy, dismemberment, swallowing up by another body—all these acts are performed on the confines of the body and the outer world, or on the confines of the old and new body. (1984, 316)

The proliferation of surface reasserts not objectivity but materiality. The body becomes part of what Bakhtin calls the anatomical and physiological series (1985, 171), not the abstinent abstraction conceived by science out of bourgeois propriety but a Rabelaisian body inflected with death, sex, defecation, ingestion, parturition. The brute physical, physiological, sexual, scatological body is infused with crude vitality. The access of materiality is also a perversion of individuality. Adumbrating surface inscriptions to grotesquery make the body a fabulous monster. Subjectivity is as apt to slip off of these hypertrophied marks of individuation as to cling on to them.

The body bears traces, signs, inscriptions of its ontological status. Inscriptions are taken to disclose the lineaments of subjectivity on the body. Incisions are surgical inscriptions. They mimic the body's own capacity to reify, perforate, and proliferate surface. Such inscriptions, incised, like tattoos, into the body do not *belong* to the body; they are not owned, as property; they are incorporated (Stewart 1984, 27). The body sports cameos and intaglios as features of its corporeality. At the same time, surgery inscribes another discourse on the body. Cuts and scars are writings on the body. Writings, as Stewart notes (1984, 14), are "tracks of the body," like other bodily aftermaths: footprints, fingerprints, scent trails, aftertastes, what Mary Douglas calls bodily exuviae (1970, 142), the sound of footsteps in the night. All are the depressions and ridges, the residues, the impressions, left on the world by the body. Scars are the body tracks of the surgeon on the patient, the trail of one body's passage across and into the terrain of another. They inscribe the body into the discourse of medicine, not effacing but adumbrating its instantiation as a social body. Surgery as a discourse overwrites the body's inscription as a cultural text. Both inscriptions materialize as characters of a script into which we read personhood. So bodily inscriptions affect our deciphering of signs of presence in the flesh.

The body is supposed at once to conceal and to express an inner self, as if the self were secreted inside the body but could somehow be exuded out of it. Materializing a self on the surface of the body reduces access to what might be regarded as the real but interiorized self. "In short," Goffman writes, "since the reality that the individual is concerned with is unperceivable at the moment, appearances must be relied on in its stead. And, paradoxically, the more the individual is concerned with the reality that is not available to perception, the more he must concentrate his attention on appearances" (1959, 249). Hence the exigency of our inspection of the surface of the body for signs of presence in the flesh. The skin suggests itself as the integument on which the self is written as well as the one in which it is suspended.

MAKING PATIENTS

When the surgeon comes into the operating room, the patient is already laid out on the table, asleep and naked. It is into that bare skin that the surgeon cuts. What sort of intimacy is thus transacted? With what delicacy and bluntness? With what modes of attention and abstraction? Under what forms of representation? And, especially, between whom? How is subjectivity—and objectivity—constituted here? Who is present in the flesh? And so, what transpires?

Typically, perception clusters on the surface of the body, the site where self and world intersect. It is because the perceptual apparatus is surfaced in this way, Drew Leder holds, that I can convey and capture information corporeally (1990, 11). But in the act of perception, my surface absents itself. I reconstitute myself at the focus of my perceptual objects, my body becomes, as Leder puts it, ecstatic, and, as I do so, my body withdraws from my awareness, recedes, disappears, and becomes, in Leder's term, recessive. "I inhabit one body, ecstatic-recessive in its entirety. Each organ both projects outward and recedes inward, eluding the self bidirectionally" (1990, 56).

Visceral processes are instances of this "corporeal disappearance" (Leder 1990, 36). My interior is radically inaccessible to me. I am hard-put to influence by conscious intent my digestion or my heartbeat or my brain waves. I am aware, episodically, of visceral sensations but "such perceptions are highly limited, traces of a vast invisible realm" (1990, 61). Leder describes the "ineluctable discontinuities" of these visceral processes. "But these are intermittent punctuations in a shroud of absence" (1990, 42).

Because of the recessiveness of the viscera, influencing visceral processes requires remarkable expertise. Medicine, along with such practices as meditation or psychology, is such an expertise. "The need for highly specialized training is evidence of the perceptual reticence of our viscera as compared to the body surface" (Leder 1990, 43). The never-effaced ambiguities of diagnosis are artifacts of "visceral disappearance" (1990, 51). It is this withdrawal of the viscera, their inaccessibility to our introspective attention that permits the objectification of the body. The reticent interior characteristically deflects self-awareness. Recovering a phenomenology of the interior of the body, even as the viscera recede from apprehension, disrupts the dualism which situates the medical body in the realm of objectivity.

As the surgeon approaches, the anesthetized patient is experiencing "a mode of depth disappearance now enfolding the body in its entirety" (Leder 1990, 57). Sleep withdraws from me my capacity to act *from* my body and the capacity of anything else to act *toward* my body (terms from Polanyi 1969, 138–158; in Leder 1990, 15). My body is altogether absent from me. Anesthesia ensures this absence during surgery. As the patient draws away into depth disappearance, the surgeon is witness to a transformation: the body is drained of subjectivity. This can be experienced as an evanescence and subsequent recrudescence of personhood. Herbert Graham, lightly sedated but not anesthetized, is awaiting surgery in the operating room. The anesthesiologist, Simonetta Fiori, inserting a needle for the anesthetic into the crook of his arm, addresses him as "Mr. Graham." As he drifts off, she shifts to "Herbert." By the time the surgeon, Adam Tartakoff, comes in to consult about the anesthetic, Dr. Fiori is referring to the patient as "buddy." As she administers the anesthetic and his body is enfolded in depth disappearance, she calls him "Herb." Once the anesthesia takes hold, the patient ceases to be addressed as a person. Four hours later, as Mr. Graham comes out of anesthesia, the process is reversed. The resident, Dr. Chavez, calls out to the barely conscious patient, "Herb, how're you doing?," and then answers himself, "Herb's doing just fine." After surgery, the patient is transferred to the surgical intensive care unit. As he hovers near consciousness, the anesthesiologist, who has accompanied him, and the attending nurses, address him as "Herb" or "Herbert," returning to formal address only as he returns to full consciousness (July 2, 1986, aneurysm description, Young). The increased use of diminutives accompanies the diminution of personhood, as if the patient initially became, not an object, but a lesser subject, on the order of a child. Herbert Graham, when present to his body, is ninety-two years old.

Unconsciousness appears to complete the transformation of the body from subject into object. When I am alert, the gaze of the Other thematizes my body for me, makes me aware of my own substantiality, my particular incarnation, the flukes and congruencies of my person. I am projected into my own objecthood. I can permit this inspection by lowering my eyes, gazing into the distance, falling into an abstraction. In so doing, I allow myself, for the nonce, to become the object of the Other's scrutiny. But I can appropriate these same devices to reembody myself by looking up, looking back, looking *at* the Other, and so insisting on my presence in the flesh. If the averted eye, the rapt gaze, the abstracted mood, *permit* looking at the body as an object, unconsciousness *invites* it. Profoundly anesthetized, I am unresistingly available to the physician's objectification.

When the body is unconscious, its sheer materiality becomes prominent. Before surgery, the patient is often anesthetized in an anteroom and then rolled down the hall on a litter to the operating room. There the body is transferred from the litter to the operating table. The surfaces are the same height and the litter is locked into position. Nonetheless, the transfer can be troublesome. Anita Brown is giving one of her kidneys to a relative. She happens to be huge, a mountain of flesh on the litter. Two of the residents work their hands in under her middle, one on each side. Someone takes the head. I am assigned the feet. The residents count, grunt, and heave her up. Her skin feels warm, dry, and preternaturally smooth. Am I simply unaware, under this circumstance, of the ordinary anomalies of skin surface? Her feet are surprisingly light but somewhat awkwardly coordinated with her torso. I wrap my arm around her ankles. Better to embrace her bodily than to cradle her heels fastidiously in my palms and lose control. We lift; she is over (1979, transplant description, Young).

The objecthood of the body is substantiated by its subsequent handling. To get at the kidney, the body has to be tilted to one side. The surgical team heaves Ms. Brown over onto her right side, lifting up the left aspect of the chest and abdomen. When they get the body balanced, the medical student slaps her on her bare rump. The resident remarks that she does not seem to be in the right position but Dr. Tartakoff says, "I like it this way," so they tape her down. Then he asks for a kidney bar and table is arched from underneath to extend the exposure of the uplifted left side.

Cushions are tucked in around the body to keep it balanced. On reflection, Dr. Tartakoff decides she is not in the right position. "I think we have her too much over to this side." They adjust her. "O.K., you want to tip her back toward me a little bit. Perfect. O.K., that'll have to do." The resident's rump slap is the nonverbal equivalent of such aesthetic evaluations as "I like it like that" or "Perfect" (1979, transplant description, Young). They mark the shift of the body from subject to object and, in the same gesture, from sacred to profane. "It is," Goffman writes, "as if the body were a sacred object, regardless of the socioeconomic character of its possessor, but in this case the consideration given is rational as well as ritual. As might be expected, then, before and after the operation proper there can be observed minor acts of desacralization, whereby the patient is reduced to more nearly profane status. At the beginning of the task the surgeon may beat a tattoo on the leg of the anesthetized patient, and at the end he may irreverently pat the patient on the bottom, commenting that he is now better than new" (1961, 126). Profane attitudes instantiate the body as object.

Once the body is positioned, the surgeon and his team reconstitute it as a medical object. Fat people are harder to operate on than thin ones. The layer of fat just under the skin is itself delicate: it bleeds easily and is difficult to sew up. But the width of the fat layer also increases the depth of the hole in which the surgical team must operate, through which they must pass to get to the kidney. On Ms. Brown, this layer is a handspan across. But the fat also greases the instruments, the gloves, the suture material, and the vessels, making fine vascular procedures like this one more difficult to perform. In the course of surgery, Henry Scott, the surgeon who is opening up the recipient in the adjoining operating room at the same time that Adam Tartakoff is taking the kidney out of the donor, comes in to see how far along his colleague has gotten. Dr. Tartakoff says, "Hi, Henry. This lady had a one-legged sea captain strapped to her side and we had just a little trouble getting him off." Dr. Scott replies, "That is a gigantic hole. What is that up there, the heart?" "I guess so," says Dr. Tartakoff. "See any cats in there?" says Dr. Scott (1979, transplant description, Young). Allusion, however witty, to the woman as a whale attests to her objectification. In the first instance, it emphasizes her solidity. She is reduced to her heft. In the second, it elides her personhood. Such a remark would never be made in her presence. It presumes an absence.

Dr. Scott's response, "That is a gigantic hole. What is that up there, the heart?" shifts attention to the body as object, specifically as the site of a cleavage, an excavation. Its teasing inflection acknowledges the competitiveness between these two surgeons. Dr. Tartakoff is reputed to make big incisions but to operate fast; Dr. Scott is reputed to make small incisions but to operate slowly. The incision in question is, of course, nowhere near the heart. Dr. Scott has license to make this remark because his status is commensurate with his colleague's.

Dr. Scott's closing remark is also a comment on the size of the incision and at the same time a recognition of the impropriety of the objectification the comment assumes. The remark alludes to a story about one of the old master surgeons, reputed to be a brilliant technician but without tact or sensitivity. The surgeon is about to operate on a nine-year-old child and her father asks the surgeon how big the incision will be. "As big as it has to." "Could you give me some idea?" asks the father. "Well," says the surgeon, "it'll be just about big enough to drop a cat through" (1979, pers. comm.). The allusion is designed so that only hearers who already know the story will catch it. Stories like this are kept as a secret lore among surgeons because the profession of surgery is felt to be vulnerable to criticisms of just this kind. Ironically, it is their sensitivity, not only to the criticism but also to the embodied person, that prompts their tact here. It is nonetheless the violation of subjectivity that amuses them. Black humor rides on the profanation of the sacred, whose boundaries it thereby locates.

The subjectivity of the patient is further impaired by the removal of clothes. Clothes are one of our self-inscriptions, our articulations of ourselves into social space. Nakedness contracts us to our skin. Our social presences are dismantled with our clothes. Naked, I become acutely aware of being in my skin. It is not so much that my skin fails to conceal me as that it fails to reveal me, to put forward my cultivated social encrustations. What I am bare of, as Goffman would say, is a carefully staged front (1959, 22, 252–254). As long as I undertake these divestments myself, I can still infuse my subjectivity into my corporeal presentations. As intentionality passes from me, so, too, does my subjectivity.

Ruth Anderson is what is called an emotionally related donor: she is giving a kidney to her husband, who will be operated on in the adjoining room. In her hospital room, Mrs. Anderson takes off her own clothes and puts on a clean light green knee-length cotton gown, worn so that it opens at the back. With this on, she is wheeled to the operating room where she is assisted up onto the operating table and lies down on her back. The anesthesiologist helps her slip the sleeves of the gown off, leaving the garment lying loosely over her supine body like a blanket. Her arms are then stretched out from the shoulder onto a crosspiece extending out sideways from the operating table, taped down, and pierced with intravenous lines. Once the anesthesia is injected and the patient loses consciousness, the anesthesiologist lifts off the gown, leaving the sleeping body naked (May 21, 1987, transplant description, Young). Thus are her ever slighter vestimentary allusions to social propriety peeled away. The body, anesthetized and denuded, has already lent itself to its own rewriting. When Dr. Tartakoff comes into the operating room to start the surgery, Mrs. Anderson is already voided of her subjectivity. Her skin presents its surface to be inscribed.

Dr. Tartakoff swabs the upper aspect of the torso with an iodine scrub, starting at the proposed site of the initial incision, just below the ribs on the left side, and circling outward. Then he lays a bluish sterile drape across the lower abdomen, unfolding it down over the feet, and another across the lower chest, uplifting its upper edge to clip onto an overhead frame suspended at the level of the body's neck. A third and fourth drape are laid across the right aspect of the abdomen and the left aspect of the back and clipped to the first pair to isolate a square patch of skin. By these arrangements, the peripheral aspects of the body, including here the lower body, the limbs, the back of the body, the shoulders and the head, are excluded from the scene of operation. What is instantiated is the innermost ritual space of surgery: the sterile field.

The sterile field is inscribed on the square patch of skin. It mounts up over the drapes on the body and extends toward the foot of the operating table to include the table of sterile instruments that lies athwart it and toward the head to the curtain between the patient's body and her head. The field rises above the body of the patient in the shape of a canopy attached to the handles of the overhead lights, which are cuffed with sterile covers. It arches out over the bodies of the surgical team on either side of the operating table so that their fronts serve as its sides, and tapers out at the level of their ankles. At the skin, along the drapes over the contours of the body, under and around the instruments, over the scrub suits of the surgical team and on the handles of the lights, the field is defined by palpable boundaries. Above the heads and between the bodies of the surgical team, the realm is defined by impalpable boundaries. Along their backs, around their feet, and across the surface of the overhead lights, where the sterile field is cinctured off from its clean but nonsterile surround, realm status is especially vulnerable. Once I brushed the back of the sleeve of a surgeon's gown with the corner of my notebook and he turned around and snapped, "Watch out for the sterile field." The boundaries of the sterile field, palpable and impalpable, are clear to its inhabitants. Along its rim, the field pouches in, as it were, refitting its boundaries around the bodies of the surgical team so that their costumes and accoutrements serve as its finely articulated edges. Surgeons never enter the sterile field in their own bodies. Rather, the boundary of the space is rendered a flexible membrane which they poke in and which closes around them so that they can manipulate what is inside.

MAKING SURGEONS

If, in order to enter the sterile field, the patient undergoes a corporeal divestment, the surgeon undertakes a corporeal envelopment. Medicine is itself an enclave in the ordinary. Surgery as a field of medicine has a separate history. Originally, physicians, who deciphered external body signs and administered substances externally, were associated with alchemists and philosophers in contrast to chirurgeons who, as wielders of the knife, were associated with barbers. Only at the time of the French Revolution in the late eighteenth century, in which severing limbs surgically saved soldiers from wound infection, did the society of physicians in France agree to recognize surgeons as fellow practitioners (Stafford 1991, 53). Still, surgeons refer disparagingly to internists as fleas and internists refer disparagingly to surgeons as plumbers, cutters, or blades, this last term catching some of the swashbuckling character attributed to them. An old adage goes:

> A surgeon has
> the eye of an eagle,
> the heart of a lion
> and the hand of a woman.

And, to a version of this scratched as graffiti on the wall of the surgeon's bathroom, some disgruntled wag adds,

> the tact of a warthog.

Surgery is maintained as a separate preserve within the hospital. Operating rooms are clustered off a hall on a floor to which access is restricted. At one end of this hall are changing rooms. Surgeons, entering this preserve, peel off the outer layers of their office clothes and put on a surgical costume. The surgical scrub suit consists of a clean, wrinkled, short-sleeved, aquamarine tunic tucked into loose trousers, tied at the waist.[2] Thus the surgeon is enveloped in his or her own ontological status. This bodily envelope constitutes a portable preserve of the sort Goffman calls a "sheath" (1972, 38). The ontological status achieved in this way can be exported to the surgeon's consultations with relatives of patients or even to office visits. Sometimes on these occasions, the scrub suit is covered with a jacket that retrieves an aspect of the social person in its formal guise. The custom of wearing white coats or tunics in the office or examining room is an attempt to import into the nonsurgical realm some remnant of its mystery, albeit a spurious one. The white coats are laboratory costumes, not surgical ones. In practice in this hospital, surgeons do not wear white coats at all. Unlike members of other specialties in the hospital, they follow the custom of their predecessor, one of the grand old men of surgery, and wear conventional jackets and ties to see patients. The clout of surgery nonetheless so infuses their surgical garb that tunics tend to get filched by the unentitled. To prevent this, the hospital, at one stage, produced them in hot pink, a much-mocked and in any case ineffectual practice that was consequently short-lived.

The operating suite in the University Hospital consists of a short broad hall that opens out on one side into an alcove containing sinks and shallow embrasures with paper caps, masks, and shoe covers. The alcove has windows onto the two adjoining operating rooms that are connected by a narrow passageway at the back. When Dr. Tartakoff arrives in his scrub suit, he picks up a cap and ties it on behind his head.[3] He takes a mask and bends the rim over the bridge of his nose, unfolds its pleats over his mouth, tucks the lower rim under his chin and ties it behind his head, too. Then he slips covers, not unlike shower caps, over his shoes. For the kidney transplant he is coming to do, Dr. Tartakoff uses a head-lamp, mounted on his forehead and secured by a band around the crown of his head from which a long thick electrical cable runs down his back to the level of his calves. The lamp is powered from a wheeled generator the size of a small refrigerator to which this cable will be attached. Thus accoutred, Dr. Tartakoff goes into the operating room to position his unconscious and unclothed patient for surgery.

After securing the patient in position, Dr. Tartakoff returns to the alcove, washes his hands and forearms with a surgical scrub for five minutes, turns off the faucet with his elbows, and goes into the operating room, opening the swinging door with his hip or shoulder and holding his hands fingertips up so that contaminated water cannot drip onto them from his upper arms. Two nurses are in attendance in the operating room, a scrub nurse, who is herself already scrubbed and in a sterile gown and gloves, and a circulating nurse who is technically not sterile and can therefore fetch things from outside the operating room or transport things around its periphery. A complex choreography designed to preserve the sterile field then ensues. The scrub nurse opens a folded square of light green fabric lying on the sterile table from which she hands the surgeon a cloth to dry his hands. He does so and then tosses the cloth into a waste receptacle. The nurse takes up a long-sleeved gown and shakes it out in front of the surgeon so that he can walk into it, leaving the opening at the back. The circulating nurse comes round and ties the gown at the back. The scrub nurse then snaps a sterile flap closed over the ties where they touch the surgeon's nonsterile scrub suit in the back. The circulating nurse brings over a sealed package of gloves which she breaks open from underneath. The scrub nurse lifts out a glove and holds it by the cuff while the surgeon pushes in his left hand. Then they both take hold of the cuff of the other glove and pull it onto his right hand. The surgeon adjusts the gloves by wriggling his fingers or folding them together. Dr. Tartakoff rubs his gloved fingers down the front of his sterile gown to work them onto his fingers (May 21, 1987, transplant description, Young). The handles of the massive overhead lights have sterile cuffs so that the surgeons can adjust them when so gloved.

The surgeon's body is untransformed by these rituals. Rather, it has been enveloped in the accoutrements of another reality, the realm of surgery. This enclosing membrane remains pervious in spots, the forehead and eyes, the ears, the neck and hairline, the calves, ankles, and feet. The foot covers, like the scrub suit, are clean but not sterile. To preserve the boundary between realms, the surgeon may ask the circulating nurse, who is outside the sterile membrane, to reach in and wipe his forehead, clean his glasses, or even scratch his nose, so that these activities do not contaminate him. If the field is accidentally threatened, for instance when one surgeon cuts another's finger during surgery, their first concern is not the treatment of the cut hand but the preservation of the sterile field: the surgeon snatches out his hand and has a second glove put on over the first to keep his blood out of the operating site. The rim of the field has been attached around the bodies of the surgeons so that the outer surfaces of their garments constitute the inner surface of the sterile field. Work on the sterile field does not, in principle, penetrate into it.

MAKING OBJECTS

Dr. Tartakoff and his resident, Dr. Fields, press together, as it were, the sides of the sterile membrane to palpate the ribs and abdomen of Ms. Brown's body in order to plan the angle of the incision. Characteristically, they do not discuss skin in terms of what is concealed underneath it but in terms of what protrudes through it: the visible rib, the palpable liver. Then Dr. Tartakoff takes an electric cautery needle and incises a fine cut just below the middle of the rib cage and along the lower edge of the left rib, angling toward the back. Delicately, he cuts through slender layers of tissue, incising, carving, excavating the deep narrow passageway to the kidneys (May 21, 1987, transplant description, Young). The duration of surgery is reckoned "skin-to-skin," from the time of the opening incision to the moment of its closure, the span of its deformation of the boundaries of the body. The incision does not properly encroach on the body. Rather, it extends the lining of the body to the membrane enclosing the sterile field. The inner organs come to inhabit an expanded, partially transparent space in which they can be lifted, separated, turned or cut. The outer body of the patient, like the body of the surgeon, is excluded from this realm. Phenomenologically speaking, nobody is there.

Surgery does not threaten the integrity of the body but the coherence of our discourse about the body as a habitus of the self. We read skin as the boundary between self and world. Surgery shifts the boundary from the skin to the sterile membrane. The sterile membrane is not inscribed with the lineaments of individuality, traces of presence, so we cannot move from the perception of individuality to the apprehension of subjectivity. We cannot decipher off the membrane signs of presence in the flesh. The membrane, which is surgically designed to prevent contamination, is phenomenologically designed to preserve discretion between the realm of the ordinary in which bodies harbor selves and the realm of surgery in which they harbor something quite else. Skin is rewritten as a sort of object, an organ among other organs, part of an interior landscape. Cutting the skin does not breach the boundary between self and world because that boundary has already been dislocated by the discourse of surgery. What is revealed by the cut? Not an inner self. Merleau-Ponty writes, "I know quite well that back there is only 'darkness crammed with organs' " (1973, 133–134). We preserve our way of talking about the body as a habitus of the self by never looking in.

This dislocation of the self from the body is crucial to the humane practice of surgery. It ensures that surgeons do not commit trespasses against persons but perform operations on objects. It is akin to the occasion in *Through the Looking Glass* on which the Red Queen introduces Alice to the leg of mutton, which gets up and bows. Alice takes up a knife and fork to offer the Queen a slice and she says, "Certainly not, it isn't etiquette to cut anyone you've been introduced to. Remove the joint!" (Carroll 1989, 296).

Thus, throughout the surgery, the recessive body of the patient beguiles the ecstatic body of the surgeon. The surgeon, excluded bodily from the sterile field, is nonetheless passionately invested in it. The workings of his own body recede from him, and he enters instead into the recesses of the Other. Surgery is a foray into alien terrain.

> Turn sideways, if you will, and slip with me into the cleft I have made. Do not fear the yellow meadows of fat, the red that sweats and tickles where you step. Here, give me your hand. Lower between the beefy cliffs. Now rest a bit upon the peritoneum. All at once, gleaming, the membrane parts . . . and you are *in.*
> (Selzer 1976, 25)

But it is his own subjectivity the surgeon realizes there; the Other's eludes him.

The surgeon has come into the patient without his body, out of his body, but only because his body releases him, disappears. Occasionally other members of the surgical team come, too. Their bodies, like his own, pass from his awareness. He and they are jointly projected into a space and time beyond their corporeal boundaries. They experience themselves not so much as in synchrony as out of body. It is as if their intention accomplishes itself. This effect has been called a "flow experience"; the surgeons themselves call it "mo," for momentum. Only when difficulties arise, when there is a hitch in the proceedings, does the body, the surgeon's or another's, obtrude itself.

Ms. Brown's kidney has been taken out. The surgeon and his associates are now reassembling her interior organs to accommodate its absence. The assemblage must allow her second kidney to take over the function of both. The surgeons have a special feeling about this patient. Unlike the recipient, the kidney donor comes in perfectly healthy. They would like, as they say, to send her out no worse than she came in. Hers, after all, is the altruistic gesture. Something prompts the resident to remark, "I hate to say it, if we lost that artery—." Dr. Tartakoff replies, "I'd rather lose the artery than the vein. The vein is friable and it tears easily when you pull it back; the artery is tougher, you lose a lot of blood—JESUS ((blood spurts over the surgeon's mask and glasses)). All right. Everybody hold. Don't do anything. Give me a right angle. Hold that. O.K. Pull it . . ." He speaks to the anesthesiologist, "We lost the artery. You better get some blood in here." "We'll do everything we can for you," replies the anesthesiologist (1979, transplant description, Young).

Only moments after he mentions it, the artery has slipped out of the instrument with which the resident was holding it and retracted into the viscera, sprung free like an elastic band. Dr. Tartakoff, fingering in among the organs to locate its end, is worrying about the patient's survival with blood pumping out of her. "Come on, Anita." And to his resident, uneasily shifting organs out of the way in the artery quest, "O.K., attaboy, O.K., good." He catches hold of the end of the vessel, slippery and bleeding. "I still don't have it a hundred percent. Hang loose." And to the anesthesiologist, "How's she doing?" "Fine. Stable." He pulls the artery out from behind the viscera and then turns to the circulating nurse and says, "You'll have to clean my glasses." The circulating nurse takes them off, wipes them, and slips them back on. Dr. Tartakoff says, "Over my ear. I think you don't have it quite worked out." "You have too much hair, Dr. Tartakoff," she replies. This teasing interchange signals that the crisis is over.

As Goffman points out, the surgeon may modulate and thereby establish and control, and establish control over, the rhythms of the operation by marking coming out of a crucial or tense phase by making a little joke, teasing, or mocking (1961, 126). The technical standards of the operation are the surgeon's responsibility, and he is in a position to criticize the staff. Doing so ruthlessly, however, may fluster them so that they commit more and worse technical errors. Thus a fluent modulation from sharp retorts ("If you don't have it, get it") or even fits of rage in which instruments are thrown, to teases ("Watch it," when the circulating nurse adjusts the back of the surgeon's gown), jokelets (calling the team "kids") and self-mocking displays (clipping and flipping hemostats, tiny clips for compressing a bleeding vessel, with the "verve and control that parking attendants manifest while parking cars") (Goffman 1961, 127) can serve to sharpen up and smooth down feelings over the course of a procedure. In this instance, a resident's slip of the hand suddenly endangers the life of the patient. Bodies have malfunctioned and thereby obtruded themselves on the surgeon's attention. Dr. Tartakoff freezes everyone in his or her tracks, moves them, loosens them up, tightens them down. Their bodies, the patient's, the resident's, his own, have suddenly become obdurate and incalculable. They have, as Leder puts it, "dys-appeared." The surgeon and his team are precipitated into their own bodies and the ecstasy leaves them. Gingerly, fully corporealized, they approach again the opened body of the patient to lose themselves in its interior.

In this ecstatic grasp of the visceral recesses by the surgeon, the patient "utilizes the body of another as an indirect conduit of perception and control" (Leder 1990, 51). It is I who hunt myself there. The surgeon acts as my surrogate, manipulating inner objects to affect my states of being. This complicity colors our relationship. It is as if we are both in on something, something he never says but I nonetheless wait to hear. Through his descriptions of procedures, I cock an ear for this disclosure, this hint of myself which he might yield up. Yet this surgical exposure of the viscera does not overcome the inaccessibility to me of my own interior. Not even if I look myself.

> He is not asleep, but rather stares straight upward, his attention riveted, a look of terrible discovery, of wonder upon his face. Watch him. This man is violating a taboo. I follow his gaze upward, and see in the great

> operating lamp suspended above his belly the reflection of his viscera. There is the liver, dark and turgid above, there the loops of his bowel winding slow, there his blood runs extravagantly. It is that which he sees and studies with so much horror and fascination. Something primordial in him has been aroused—a fright, a longing. I feel it, too, and quickly bend above his open body to shield it from his view. How dare he look within the Ark! Cover his eyes! But it is too late; he has already *seen;* that which no man should; he has trespassed. (Selzer 1976, 25)

The viscera as source of inner sensations remain as opaque to the surgeon as they are to me. What I see is a exotic landscape.

Surgery constitutes a defamiliarization of the body. The discourse of anatomy inscribes the body as an assemblage of objects. In the course of surgery, Dr. Tartakoff quizzes his intern, Dr. Stevens.

Dr. T: What's right here.
What hollow viscus is right here is that—
Dr. S: Oh
duodenum?
Dr. T: Right.
(Transplant, May 21, 1987, Lefkowitz transcript, 10)

It is because these organs seem foreign to me that I have no sense of privacy about my interior. Anyone may see intimate photographs of my bones, televised images of my circulation, even my inner organs themselves. They are foreign objects, not-self.

Under the same logic, the discourse of surgery reinscribes deformations of the flesh as operations. Dr. Tartakoff says to his resident, Dr. Fields:

Dr. T: And then you close up.
Dr. F: All right.
Dr. T: Double-layer closure
internal oblique and transversalis
and a (running)
heavy stitch
whatever you like
and interrupt it
in the external oblique.

Dr. F: Absorbable sutures?
Dr. T: Yeah.

(Transplant, May 21, 1987, Lefkowitz transcript, 26)

Here the body has vanished like the Cheshire cat leaving its surgical surface inscriptions visible in the form of stitches.

Inner organs retain an ontological ambiguity only if one of the discourses of the surface is invested in them, seeing the heart as the seat of the passions, for instance, or the brain as the locus of thought. Such states of being remain invisible in these organs, however. The pressure of medicine toward materialization is an effort to reify the ethereal: to locate feeling, sensation, thought, inner states, in the substance of the body.

MAKING SUBJECTS

The surgical rewriting of the body as an object is fragile. When in the course of harvesting her kidney Dr. Tartakoff cuts through the outer layer of Mrs. Anderson's skin with an electric cautery needle, the incision is accompanied by the sound of sizzling and the singe of flesh. He says,

Dr. T: Uh she's mov—
she's feeling some of this I think.
Is it possible?
Maybe it was just a uh
a respiration or something.
Maybe not.

(Transplant, May 21, 1987, Lefkowitz transcript, 1)

The subject threatens to break through the object. Later, the anesthesiologists discuss the difficulty of keeping the patient's right arm, with the intravenous needle in it, properly positioned with a surgeon leaning on her collar bone. The surgeon, Dr. Stevens, turns to her resident and says,

Dr. S: What's the matter.
Am I hurting (her)?
Dr. F: Well you— she will hurt post-op.

(Transplant, May 21, 1987, Lefkowitz transcript, 6)

Leaning on the body, putting instruments down on it, and otherwise acting as if there were nobody there are common in surgery but, as here, commonly vulnerable. Subjectivity slips through the rewriting of the body as an object: witness, in these instances, the recurrence of personal pronouns. It is to prevent such slippage that access to surgery is restricted. If, for us, surgery threatens a discourse about the body, for the surgeons, we do. Outsiders import other readings of the body, opening up the acts of surgery to imputations of sacrilege. Notice that the only uninitiated participant in the surgery is out cold. The body is a palimpsest: its ordinary inscriptions show through the overwriting, as indeed the realm of the ordinary always presses against departures from its conventions.

These metaphysical dislocations conjure up the mind/body problem. Where is the self if not immanent in the body? What is the lining of the body if not the subject? How is subjectivity constituted in the body of the patient by the surgeon? The body, its processes, and the organs themselves are regarded as aesthetic objects. On another occasion Dr. Tartakoff's team has laid bare the kidney of a patient. It is still *in situ* but detached from everything except its blood supply. The ureter lies curled across the body of the organ like a worm. Surgery is in abeyance until the operating team next door is ready to implant the kidney in the recipient. As they watch, urine squirts spasmodically out of the clipped ureter. Dr. Tartakoff says, "Let's watch this a minute, it's beautiful. The ureter has its own peristaltic action." "If there were a bird, he would eat this right now," the resident responds, "There it goes. THERE SHE BLOWS." Urine flows out of the ureter in a small arc. Dr. Tartakoff says, "Has a pattern, doesn't it? It does three little spits and a big squirt" (1979, transplant description, Young). On observing this same phenomenon in Mrs. Anderson's kidney, Dr. Stevens calls out,

Dr. S: Oh there.
Look at that.

And his colleague remarks,

Dr. T: You can see— you've seen this
phenomenon of the urine coming out of the ureter before I think.
I'll show it to you
it's very exciting.

To which Dr. Stevens replies, perfectly seriously,

Oh, isn't that cute.

And Dr. Tartakoff agrees,

It's very pretty.

(Transplant, May 21, 1987, Lefkowitz transcript, 23–24)

The kidney achieves the status of what Robert Plant Armstrong calls an "affecting presence" (1975, 21, 42–70; 1971): the notion that aesthetic objects achieve a resonance beyond their objecthood.

Indeed the surgical project can be conceived as an aesthetic one. Dr. Tartakoff and his resident, Dr. Ryegate, are performing an amputation of a foot and lower leg. They debate which sections of muscle need to be trimmed away on account of injury and which need to be preserved for the reconstituted limb.

Dr. R: This has to go
[[[ ]
Dr. T: So this has to come out.
Some of this has ( )
[[[ ]
Dr. R: ( ) away.
Yes.
Dr. T: O.K.
Let's take some of that out.
[[[ ]
Dr. R: How about if I
trim this with the amputation knife.
Dr. T: O.K.
Cut it longer than shorter.
I mean you'd rather uh
you can always go back.
Watch the skin.
/
((To scrub nurse)) Kocher *((an instrument)).*
/
(So), I just am afraid
if we trim off too much
we're trimming off collateral flow

who knows where it's coming from.
((To scrub nurse)) Hemostat.
That looks lovely.

[[[   ]
Dr. R: Knife back.
Huh?
Dr. T: That looks lovely.
Look at that.
/
Nice.

(Amputation, Spring 1987, Lefkowitz transcript, 29)

When they saw off the bone of the leg, Dr. Tartakoff wants to make sure the end is beveled so that tissue can be mounted up over it as a weight-bearing surface. His resident is struggling with sawing the bone. The surgeon says,

Dr. T: Almost.
You're doing good.
Push.
/
O.K.
Dr. R: Now let me
Dr. T: This is not going to be easy
this part.
Want me to try?
Want me to?
No go ahead you=
Dr. R: (do it or I could go out) ( )=
[[[   ]
Dr. T: No go ahead.
Let me do it then.
Might as well just get it off=
Dr. R: O.K.
Dr. T: Cause I got a shot.
Dr. R: This is the— this is the=
Dr. T: fun part?=

Dr. R: artistry I think=
Dr. T: O.K. go ahead (wise guy).
(Amputation, Spring 1987, Lefkowitz transcript, 17)

The surgeons conceive some of the issues in surgery in explicitly aesthetic terms. They puzzle over the proper reassemblage of remnants of tissue:

Dr. T: Y— y— you worry about when he flexes that
what's going to happen.
What sort of dys— function he's going to have.
I'm almost inclined to
not bring this over and cut it off
[[[ ]
Dr. R: —cut it off.
Dr. T: And just bring that over like that.
Dr. R: Well?
Dr. T: I— I think I'd prefer that.
This just bothers me
it's— aesthetically
and possibly functionally—
it shouldn't be there.
(Amputation, Spring 1987, Lefkowitz transcript, 46)

On this view, a defective object might lose not only its functional status but also its aesthetic one. On first seeing Ms. Brown's kidney, Dr. Tartakoff says, as if it were an aesthetic critique, "Look at that. Ovarian vein into renal vein. Never happens. O.K. we've got an anomaly here" (1979, transplant description, Young). Dismantling the aesthetic frame banishes the affecting presence.

If surgery transforms the body into an object, the surgeons then reconstitute in the object, the subject. In an extraordinary gesture of incarnation, the surgeon relocates lineaments of individuality in the interior of the body of the patient and then regards these as traces of subjectivity. When Dr. Tartakoff begins to trim out Mrs. Anderson's kidney, he says,

Dr. T: You can tell she's slightly older— How old is she in her
mid-thirties?
Or her late=

Dr. F: No.
Dr. T: thirties?
Dr. F: No.
Forty.
[[[ ]
Dr. T: Is she?
Um because the—
the area around the kidney's a little
stuck.
As patients get older I get—
they get uh some tiny little bits of
inflammatory atherosclerotic (change).

(Transplant, May 21, 1987, Lefkowitz transcript, 6)

These adhesions, lineaments of individuality, accretions of a person with a history, so signify subjectivity that even this interior landscape is haunted by the self.

Surgeons regard kidneys with a tenderness usually reserved for the impersoned.

Dr. T: Also it's interesting
respiration has an effect on it.
It's as if the— it's compressing the
kidney in some way
but you notice the urine is also made at a time when uh
it's an inspiration.

(Transplant, May 21, 1987, Lefkowitz transcript, 27)

As she breathes, so the kidney squirts.

This peristaltic activity persists even after the kidney is excised out of the body, and with it a ghost of the self. Dr. Tartakoff takes the kidney, holding it cupped lightly but firmly in his hands, over to a sterile table, inspects it, perfuses it with a fluid to cool and preserve it, and drops it into a metal bowl. He carries the bowl aloft, pinching off, as it were, a piece of the sterile membrane to enclose it, and transports it bodily along the back passageway to the adjoining operating room. His circulating nurse processes along behind him, rolling the power box for his head-lamp, attached by means of the long cord which uplifts the tail of his gown at the back. This ceremonious departure appears to honor the inspirited organ.

Paul Ricoeur suggests that texts present "reliefs" so that different topics appear at different altitudes (1971, 548–549).[4] Just so with the body. At the altitude of the skin, a self inheres in the body. The inner self is part of our discourse about the surface of the body, as if the self actually had a physical placement inside it. At other altitudes an alien topography materializes, one is which there is no self. Surgery inscribes this region with its own signs of presence in the flesh.

FOUR

• • •

STILL LIFE WITH CORPSE • PATHOLOGY

The grotesque ignores the impenetrable surface that closes and limits the body as a separate and completed phenomenon. The grotesque body image displays not only the outward but also the inner features of the body: blood, bowels, heart and other organs. The outward and inward features are often merged into one.

—*Mikhail Bakhtin*

Dr. Mercurio, the chief pathologist, escorts me through a labyrinth of corridors in the basement of the University Hospital to the morgue. We enter an antechamber and there, through an open archway, are four figures in aquamarine scrub suits, masked, capped, and gloved, and the corpse, waxy yellow, laid out on the examination table. The face is tipped toward me, eyes closed, the corners of the lips tilted up in a slight triangular smile. A tendril of black hair clings to the far shoulder. The arms are bent at the elbows so that the hands rest curled against the chest. The abdomen has been opened in a square flap, folded down like an apron covering, as it were modestly, the genitals. Viscera spill out over the rim. The body is utterly immobile, impossibly still, a wax figure. From its abdominal cavity effuse lush glossy inner organs in bruised purples, fatty yellows, membranous whites.

One figure, the technician or pathologist's assistant who performs most of the autopsy, stands behind the corpse, lifting out lengths of intestine, holding them aloft, clipping away their adhesions of fat. These internal tissues have a grey bloom. The other three figures approach us, two men and a woman. Dr. Mercurio introduces me to them, and we explain my research. The three stand before us in a semi-circle, nodding and speaking through their masks, their eyes flicking and glowing. "What did she die of?" I ask, although in fact, despite her nakedness, her gender is almost erased. The body is swollen, the thighs huge, belly mounded up, breasts flattened down. Except for the face, it is without particularity. There, her particular personhood is exquisitely inscribed. She died of liver disease; that is why she is so yellow. She has a lacy spray of blood across her lower cheek, fanning out from the mouth. The disease causes internal bleeding. With death, the rosy flush of the organs, the transparency of the skin, turn opaque. They won't ordinarily be so yellow, they say, meaning the other corpses I will see. Dr. Mercurio and I withdraw. "Her expression is so sweet," I remark. "Perhaps," he says, "it was a sweet release." Sometimes, after death, a rictus of the facial muscles draws up the corners of the mouth giving the face the semblance of a smile.[1]

THE GROTESQUE BODY: BAKHTIN

The medieval body is held by Mikhail Bakhtin to be "coarse, hawking, farting, yawning, spitting, hiccupping, noisily nose-blowing, endlessly chewing and drinking" (1985, 177). So the body is expulsed by ascetic medieval ideology as "licentious, crude, dirty and self-destructive" (1985, 171), thus inventing an opposition between the spiritual, aristocratic, or ethereal central discourse and the bodily, vulgar, or grotesque peripheral one. The central ascetic discourse is then raised, in literal as well as in symbolic space, toward heaven and the peripheral bodily discourse lowered to earth, or even into hell, so that the opposition is rendered hierarchical. The hierarchy is then reinscribed on the body, whose topography is appropriately oriented, so that the face and head are opposed to the genitals, belly and buttocks (1984, 21), the region of the body Bakhtin describes as the "material lower bodily stratum" (1984, 368). The body in general, or its lower parts in particular, come to represent the tabooed discourse in the form of what Bakhtin calls "grotesque realism" (1984, 18).[2] Peter Stallybrass and Allon White write:

> Grotesque realism images the human body as multiple, bulging, over- or under-sized, protuberant and incomplete. The openings and orifices of

> this carnival body are emphasized, not its closure and finish. It is an image of impure corporeal bulk with its orifices (mouth, flared nostrils, anus) yawning wide and its lower regions (belly, legs, feet, buttocks and genitals) given priority over its upper regions (head, 'spirit', reason). (1986, 8–9)

The grotesque body is contrasted to the classical body, modeled, Bakhtin argues, on Greek statues. "The classical statue," Stallybrass and White note, "has no openings or orifices whereas grotesque costumes and masks emphasize the gaping mouth, the protuberant belly and buttocks, the feet and the genitals" (1986, 22). By contrast to the classical body, in the grotesque body, Bakhtin writes:

> The stress is laid on those parts of the body that are open to the outside world, that is, the parts through which the world enters the body or emerges from it, or through which the body itself goes out to meet the world. This means that the emphasis is on the apertures or the convexities, on various ramifications and offshoots: the open mouth, the genital organs, the breasts, the phallus, the potbelly, the nose. (1984, 26)

Orifices and protuberances are sites where the boundaries between the body and the world are blurred: bodily substances are spilt out, stuff from the world taken in. Dissection incises fresh cuts, slits, apertures into the body. The extrusion of the lining, the disgorging of entrails over the edge of the incision, the plunging of hands and instruments into its recesses, the interchange of substances, confuses the distinction between inside and outside. When, in the course of the dissection, organs are cut loose, pulled free, and plucked out, the conventional order of the body is destroyed. Susan Stewart writes, "the grotesque presents a jumbling of this order, a dismantling and re-presentation of the body" (1984, 105) so that the body becomes, in Bakhtin's conception, a body of parts (Stewart 1984, 105). The dissected body is a grotesque body.

A portly old man with fresh skin and white hair is laid out face up on the autopsy table. His body is dappled with purplish blotches on the underside of the head, back, and legs. The lips are drawn upward and outward, slightly curled, giving the face a sweetness of expression. The chest is lifted up onto a curved block so that the head drops back. The technician sharpens a broad-bladed knife on a steel file. He cuts into the corpse at the upper inner hollow of the shoulder, runs the belly of the knife across the ribs, around the breast, along the lower edge of the rib cage, and up across the ribs to the other shoulder in a single stroke. He makes a second cut from the middle of the first down the abdomen to the pubis, skirting the navel to create a Y-shaped incision. The upper flap of tissue is detached from the ribs and draped over the chin. The two lower flaps are folded outward and clipped to the skin. Blood wells up in the abdominal cavity. He dabs it up, in an incongruously homely fashion, with paper towels.

The technician saws out sides of ribs in triangular sections, lifts them off the chest and piles them on the autopsy table. He ties off the intestines at the top, twice, close together, and cuts between the ties. He lifts loops of bowel, trimmed loose, out of the abdomen into a pan on the corpse's lap. He dissects out structures in the throat, disconnects organs in the pelvis, detaches them from the sides of the cavity and then from the top. He scoops the heart, lungs, and viscera out onto a board, leaving the clean hollow trough of the abdominal cavity, its lining grey, thick, smooth, and undifferentiated, like the skin of a dolphin. The procedure is called abdominal evisceration.

A DISCURSUS ON SAUSAGES

Now a brief discursus on sausages, and on feces, intestines, pigs, food, and phalluses. Consider first feces, "conceived as something *intermediate between earth and body*" (Bakhtin 1984, 175), what Bakhtin calls "gay matter" (1984, 335), that constitutes "part of man's awareness of his materiality, his bodily nature, closely related to the life of the earth" (1984, 224). "Dung and urine lend a bodily character to matter," he writes. "If dung is a link between body and earth . . . , urine is a link between body and sea" (Bakhtin 1984, 335).

Consider, then, intestines, that thin, tough, translucent tube which contains feces within the belly. Consider this same tubing, taken from the pig and stuffed with finely minced offal from the pig's belly (see Figure 1). Tripe, consisting of both stomach and bowels, Bakhtin describes as "the bowels, the belly, the very life of man" (1984, 162). And the pig, ambivalent animal (Stallybrass and White 1986, 44), is itself intermediate

Figure 1 Woman making sausages. (From a Flemish Psalter, first quarter of the fourteenth century. MS Douce 5, fol. 7r. Reproduced with permission of the Bodleian Library, University of Oxford.)

between categories, wild and tame, country and city, animal and man.[3] Stallybrass and White point out that "the pink pigmentation and apparent nakedness of the pig disturbingly resemble the flesh of European babies" (1986, 47). It is the animal that eats feces and makes food (Stallybrass and White 1986, 45).

Thus the mixture of ingredients according to Bakhtin:

> The bowels are related to defecation and excrement. Further, the belly does not only eat and swallow, it is also eaten, as tripe . . . Further, tripe is linked with death, with slaughter, murder, since to disembowel is to kill. Finally, it is linked with birth, for the belly generates.
>
> Thus, in the image of tripe life and death, birth, excrement, and food are all drawn together and tied in one grotesque knot; this is the center of the bodily topography in which the upper and lower stratum interpenetrate each other. (1984, 163)[4]

Hence, he continues, "the carnival role of butchers and cooks, of the carving knife, and of the minced meat for . . . sausages" (1984, 193). Its source is the dismembered body, minced flesh, "based on the grotesque image of the dissected body" (1984, 194). "Carne levare (the roasting and taking up of meat which probably gave carnival its name)," according to Stallybrass and White (1986, 184), was the festival in which "the pleasures of food were represented in the sausage and the rites of inversion were emblematized in the pig's bladder of the fool," the source, as they so sagely point out, of the balloon (1986, 53).

And the sausage, enfolded in a bun whose dough is called virgin until it is inseminated by the leaven (Conrad 1986, 52), obtrudes its final carnivalesque transformation. Carnival, ritual of contradiction, "commingling of categories" (Stallybrass and White 1986, 27), site of what Stallybrass and White call "demonization, inversion, hybridization" (1986, 56), presented by Bakhtin as "a world of topsy-turvy, of heteroglot exuberance, of ceaseless overrunning and excess where all is mixed, hybrid, ritually degraded and defiled" (1985, 8). "Eating and drinking," writes Bakhtin, "are one of the most significant manifestations of the grotesque body. The distinctive character of this body is its open unfinished nature, its interaction with the world" (1984, 281). The nexus in which sausages, feces, intestines, pigs, food, and phalluses are entangled is the belly of the dissected body. The corpse, in the act of being eviscerated, is carnivalesque.

The carnivalesque move is to turn upside down or inside out, to invert or reverse, to transgress the boundaries between discourses or to switch their content, just as dissection does the corpse. Autopsy thus evinces an affinity with other inversionary gestures like "the cartwheel, which by the continual rotation of the upper and lower parts suggests the rotation of earth and sky. This is manifested in other movements of the clown: the buttocks persistently trying to take the place of the head and the head of the buttocks" (Bakhtin 1984, 353). "Down, inside out, vice versa, upside down, such is the direction of all these movements. All of them thrust down, turn over, push headfirst, transfer top to bottom, and bottom to top, bottom in the literal sense of space and in the metaphorical meaning of the image" (Bakhtin 1984, 370). The spatial displacements of dismemberment disturb the conceptual proprieties of the body.

The grotesque body becomes a repository for degradations made flesh. Curses and abuses, obscenities and sacrileges, assault a high discourse and transform it into the low by materializing the discourse bodily, especially in the nether regions of the body. "To degrade by the grotesque method," writes Bakhtin, "they send it down to the absolute lower bodily stratum, to the zone of the genital organs, the bodily grave, in order to be destroyed" (1984, 38). Hence the association of grotesque realism with death, with the underworld, and with the demonic (Bakhtin 1984, 301). One curious indication of the corpse as carnivalesque: the morgue is always located in the basement, geographically low, constituting itself an underworld, a realm of the dead. The material lower bodily stratum is, according to Bakhtin, a carnivalized underworld (1984, 395).

To cast down a discourse, then, is not only to carnivalize it but also to materialize it, to render it matter, not inert matter but fecund, teeming, productive matter. "All that is sacred or exalted is rethought on the level of the material lower bodily stratum" (Bakhtin 1984, 370). Hence the connection, rooted in the old agricultural cycle, between death and fertility. "To degrade is to bury, to sow, and to kill simultaneously . . . , to concern oneself with the lower stratum of the body, the life of the belly and the reproductive organs; it therefore relates to acts of defecation and copulation, conception, pregnancy and birth" (Bakhtin 1984, 21). The slit belly of the corpse, exuding and engulfing matter, undergoing transformations, retains this material vitality. Death is "made out of the same stuff as life itself" (Bakhtin 1981, 195). The corpse, split and eviscerated in the morgue, remains the site of an inversionary, fertile, grotesque discourse. This carnivalesque conception of death is contrasted with the conventional view of the corpse as absence, barrenness, and stillness.

MEDICINE AS AN ARISTOCRATIC DISCOURSE

Suppose, in the face of the grotesque body, you are in the business of constructing an aristocratic discourse?[5] How do you go about it? You press the discourse through an etherealization cycle: on the one hand, raise the discourse hierarchically, purify it, sacralize its language and, on the other, exclude or suppress the low, the grotesque, the material lower bodily stratum. This is difficult in the case of medicine as it is so ineluctably a discourse of the body, in the case of surgery and pathology especially, a carnivalesque body: one split, severed, dismembered, eviscerated, turned inside out. Its material aspect appears inexpungeable.

* * *

The body of a delicately made woman lies flayed open on the autopsy table. The chest is lifted up on a block, arching the back and tilting the head back so that the top of it rests on the table, chin pointed upward, vessels prominent in the forehead. The abdominal evisceration has included dissecting out the throat structures, leaving the abdomen void and the neck a hollow stem. The technician lifts the head onto a second block, giving the macabre suggestion of a pillow. He incises the scalp from just above the right ear through the furze of grey hair to just above the left. He lifts the head and tugs the scalp down over the forehead, holding it by the hair. It folds down, inside out, over the brown face which crumples underneath like an old leather mask. He continues to cut away scalp along both edges of the incision with a scalpel. The face is now covered with everted tissue, hair inside, forming a ball of whitish integument, its protruding chin rimmed with grey furze.

The technician peels the scalp off the skull, exposing its sutures. He saws around the skull from the base of the neck to the forehead and back to the base. Bone chips fly, accompanied by the whir and grind of metal on bone and a sawdusty smell. He takes a chisel, taps into the crevice, inserts a great hook into the widened aperture, and pulls off the skull cap. The brain, in its transparent sack, droops at the back, jiggling. The technician works his fingers up between the brain and the skull at the forehead, clips it loose, then lifts the brain from underneath and scissors its deeper connections. He lowers the head and rests the brain in the bowl of the skull. Then he reaches around through the chest cavity into the hollow neck and pushes out the brain with his fingers. He removes the block under the head quickly so that the head drops back, emptying its contents like a goblet into the bowl of the skull. He sets the bowl on the table. The head, hollowed and decapitated, hangs down, the handles of its ears curving just above the rim. In the bowl sits the brain, whorled and ridged like a huge walnut.

As an aristocratic discourse, medicine becomes authoritative: monologic, monoglossic, univocal, and sacred. Authoritative discourse, according to Bakhtin, has three properties: it binds participants independently of its power of persuasion; it is both distant and high, an already acknowledged authority, what he calls a prior discourse; and "its language is a special (as it were, hieratic) language. It can be profaned. It is akin to taboo, i.e., a name that must not be taken in vain" (1981, 342). As physicians were transformed toward the end of the nineteenth century from servants who attended upon their patients into masters who were themselves attended upon, their discourse shifted toward the aristocratic. Consider Francis Delafield's 1872 description of a brain autopsy.

> The scalp is divided by an incision across the vertex from ear to ear. The flaps are directed forward and backward, taking up the temporal muscles with the skin and leaving the pericranium attached to the bone. The internal surface of the scalp and the pericranium are to be examined for ecchymoses and inflammatory lesions. A circular incision is then made with the saw and the roof of the cranium removed. The incision in front should pass through a point three and a half inches above the root of the nose, behind through the occipital protuberance. (1872, 9)

Here the body appears to disarticulate by itself: "The scalp is divided by an incision"; "the flaps are directed forward and backward." The agent of this disarticulation, the pathologist with the scalpel, has dematerialized. Its perceiver describes the events from a locus outside or above them. We, the readers, take up the same perspective. From it, the detached perceiver appears to have unlimited knowledge of that realm of events. Hence the affinity of the external perceiver with the omniscient narrator. Speaking in the present tense de-realizes the event by at once purporting to transact it right before our eyes and at the same time not doing so. The description is not of any one particular autopsy but of the course of such autopsies in general, rendering it generic rather than idiosyncratic. The particularity, the possible personhood, of the corpse is elided by the passive voice, which not only banishes agent and perceiver but also objectifies the object of perception.[6] What appears is not the act of one person on the dead body of another but a self-disclosing object (Harcourt 1987, 49–50). The dissected body reveals its affinity with sixteenth-century anatomical drawings in which the corpse is represented as holding open its own body cavities, uplifting its muscle flaps or supporting its eviscerated organs (Figure 2). In a sense, no one is there; the corpse dissects itself.

The dissected corpse is never fully embodied. It divides itself into already intelligible categories, not chunks of blood, bone, and flesh but ecchymoses and inflammatory lesions; the vertex, the pericranium, and the occipital protuberance; the temporal muscles. These Latinate terms

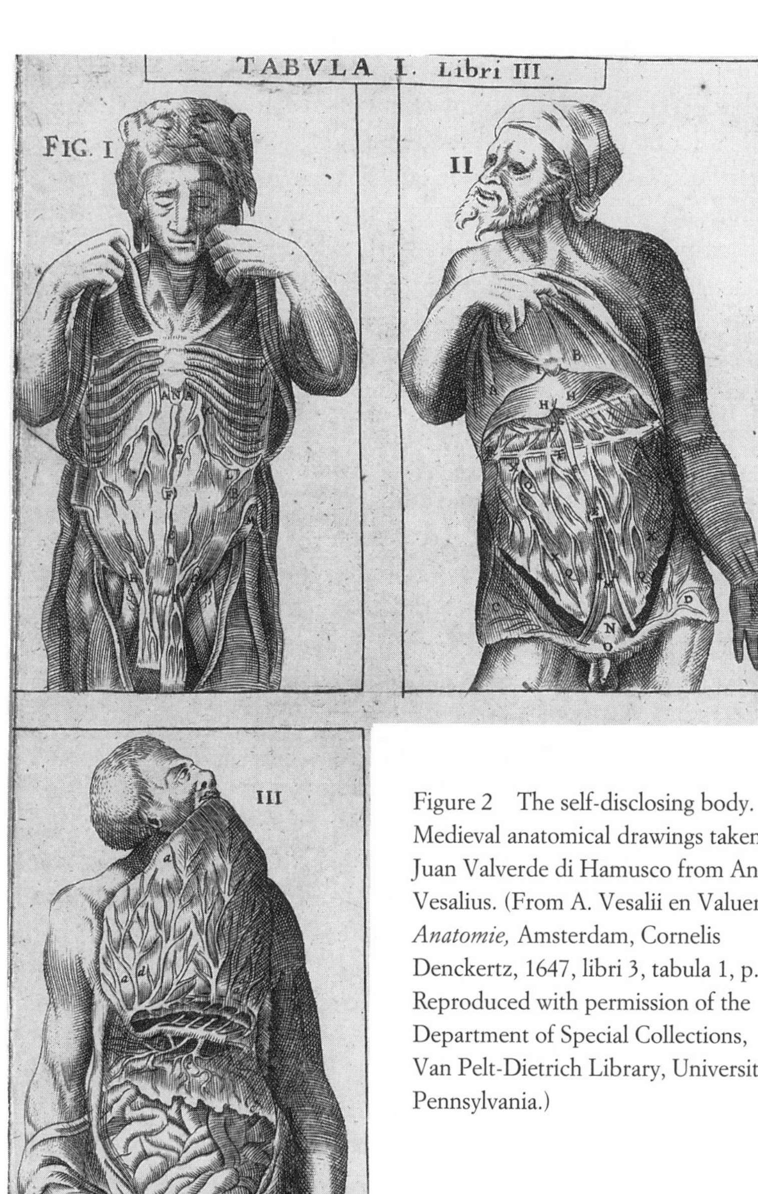

Figure 2 The self-disclosing body. Medieval anatomical drawings taken by Juan Valverde di Hamusco from Andreas Vesalius. (From A. Vesalii en Valuerda, *Anatomie,* Amsterdam, Cornelis Denckertz, 1647, libri 3, tabula 1, p. 56a. Reproduced with permission of the Department of Special Collections, Van Pelt-Dietrich Library, University of Pennsylvania.)

take their caste from association with the scientific, the academic, the intellectual, the mental. This hierarchy of the root languages of English arises, like all linguistic hierarchies, from a quirk of history, in this instance the Norman conquest of the Angles and the Saxons. As part of the aristocracy of language, Latinate terms are held to be pure as well as high, free from the vulgarity and coarseness of rude speech. And of course Latin is a sacred language whose contemporary context of use is a religious one. The Latinate terms are tinged with ritual authority.

Some contemporary texts retain the passive construction so that the body still appears to disarticulate itself but they shift toward a slightly less Latinate vocabulary and reassert the voice of the pathologist in the form of authoritative remarks, advice, or evaluation. Occasional parenthetical remarks occur in the first person (Ludwig 1979, 7, 96).

> The head is elevated slightly with a wooden block or a metal headrest attached to the autopsy table. The hair is parted with a comb along the line of the primary incision. The latter is made through an imaginary coronal plane connecting one mastoid with the other over the convexity . . . A sharp scalpel blade can then be used to cut through the whole thickness of the scalp from the outside . . .
>
> The anterior and posterior halves of the scalp are then reflected forward and backward, respectively, after short undercutting of the scalp with a sharp knife which permits grasping of the edges with the hands. The use of a dry towel draped over the scalp edges facilitates further reflection, usually without the aid of cutting instruments. If the reflection is difficult, a scalpel blade can be used to cut the loose connective tissue that lags behind the reflecting edge as the left hand continues to peel the scalp. The knife edge should be directed toward the skull and not toward the scalp. The anterior flap is reflected to a level 1 or 2 cm above the supraorbital ridge. The posterior flap is reflected down to a level just above the occipital protuberance. (Ludwig 1979, 95–96)

The pathologist with the scalpel has reappeared, if only to take instruction from the speaker as mentor. The perspective on the autopsy is now internal to that realm of events but still not embodied in its perceiver. The act of instruction is represented here. We oversee the gestures of the pathologist and overhear the instructions of the mentor but remain, as perceivers, omnipresent but disembodied, an intelligent effluvium hovering about the autopsy.

In the current autopsy manual of the University Hospital the authoritative voice of the pathologist is rendered in the imperative mode.

> Separate the hair, undoing any braids or coiffure present, to mark out the line of incision . . . Use a fresh bladed scalpel to make a single clean incision with the belly of the blade cutting down through the galea aponeurotica. With the initial incision in place, use the scalpel to cut the fascia binding the undersurface of the galea to the pericranium. With a dry towel, grasp the anterior edge of the incision and roll it forward, simultaneously cutting the strips of fascia. Turn the anterior flap down over the face to about the mid part of the forehead. In a similar fashion, turn the posterior flap to about the external occipital protuberance. (1989–90, 45–46)

Here the pathologist appears without the scalpel to issue instructions for what might materialize as an autopsy. In the instructive mode, the act remains implicit, intentional. Just as well, for here it is we who take the part of the perpetrator, who hesitate on the edge of incising the corpse. We perceive this possibility from close up. While voice remains objective, omniscient, and authoritative, perspective has shifted from the disembodied external perspective in which the perceiver, like the speaker, is both immaterial and remote from the dissection to an embodied internal perspective in which the perceiver is materialized in the morgue. It is we, the perceivers, who are embodied as subjects; the corpse takes the inscription of its objectivity from the discourse of medicine.

Dissection, shifting in the mid-seventeenth century from public spectacle to private event (Barker 1984, 13), eschewed its old association with the grotesque, the freakish, and the carnivalesque in order to present itself as refined, rarified, pure, priestly, or aristocratic. The sacralization of medicine entails the banishment or concealment of the grotesque body in the interest of producing a pure discourse. This discourse is exclusive and hierarchical with respect to ordinary discourse. The hierarchy is reinscribed within the discourse so that apprentices are themselves etherealized as they rise through the discourse of medicine to become masters. An anomaly remains. The grotesque body of pathology is a threat to the aristocratic discourse of medicine.

THE PRIVATIZATION OF THE BODY: BARKER

From the Renaissance to the modern period, the spectacular medieval body is reinvented in a series of moves exquisitely illuminated by Francis Barker. Initially the body is privatized, withdrawn from public view underneath clothes, inside of rooms, behind institutions, into writing. This privatization depends on the invention of "the public and private as strong, mutually defining, mutually exclusive categories, each describing separate terrains with distinct contents, practices and discourses" (1984, 34). Banishment dematerializes the body. It becomes, in the first instance, invisible, for, as Barker points out, "the body has certainly been among those objects which have been effectively hidden from history" (1984, 12), and in the second instance, unmentionable, for etiquette excludes not only the body but also its verbal tokens from discourse. The substitution of the terms "white" and "dark" meat for the breast and thighs of chickens is an instance of exclusion. But this "decorporealized etherealism," as Barker calls it (1984, 94), is only one aspect of the strategy. The other, and at the same time antithetical aspect, is the reinscription of the body as an object of representation (Barker 1984, 80, 97). No longer the representing subject, the body becomes Other, even to itself. Finally, the body is materialized as an object. By contrast, the self, retaining the ascetic medieval imagery, is etherealized as a subject. This ether, this essence, this absence is encoded, writes Barker, as "an individual, privatized and largely passive consciousness: systematically detached from a world which is thus beyond its grasp" (1984, 36). The ethereal subject is then interiorized in the body object. The spectacular inscription of the self on the surface of the body is erased and rewritten, inscrutably, within.

So the materiality, the objectivity, the objecthood of the body is not a natural disposition but a cultural accomplishment. The critique of materialism as idealism suggests that to see the body as material is itself an idealization (Bataille 1985, 15). The consequent incommensurability between the material substance of the body and the ethereal substance of the self constructs the mind/body problem. How is the person hitched to the body, the mind to the brain, the ethereal to the grotesque? Are mentality and physicality different orders of event? If so, as Hegel suggests, the corpse can be regarded as the ultimate presentation of body as object (1977, 114).

We arrive, via the Barker moves, at a discourse, medicine, and its object of representation, a reconstituted body. "Neither wholly present, nor wholly absent, the body," writes Barker, "is confined, ignored, exscribed from discourse, and yet remains at the edge of visibility, troubling the space from which it has been banished" (1984, 63). Consider now medical discourse and the unreconstituted body that haunts its edges.

THE SPECTACULAR, THE SPECULATIVE, AND THE SPECULAR IN THE PHENOMENOLOGY OF THE BODY: FOUCAULT

The medieval body is both homologous and analogous to the cosmos. Body and cosmos partake of the same elements, arrayed in the same hierarchy. The body is a micro-cosmos. The four elements, earth, air, fire, and water, are transmuted in the body into the four humors, melancholy, blood, choler, and phlegm. As they transmute, the humors rise and rarify so that the lower body is infused with natural spirits, the upper body with vital spirits, and the head with a distillate of pure animal spirits (Tillyard, n.d., 69). The material is thus transmuted into the ethereal.

Medieval humoral medicine is a system of elemental affinities, or sympathies, worked out according to a cosmo-logos to alleviate the dis-ease of body with cosmos. On this theory, the knowable is not necessarily visible, evident in perceptual space. Disease can be deciphered from fluid issues on the surface of the body so that it is possible to say the unseen. The interior of the body, in the process of transmuting matter, is in a state of flux, liable to disease in the form of either turbulence or induration of its fluid interior (Duden 1991, 130). Mental and material properties of the body are indistinguishable. Emotion, for instance, is a disturbance of the humoral substance. A good humor or an even temper can be restored by tempering the humors in the body so that they correspond to the elements of the cosmos. Disease and health are inherent in the humoral substance of the body. Clear into the eighteenth century, despite the desuetude of humoral theory, this "flowing and indistinctly structured body," as Barbara Duden puts it, contained "nothing solid, no scaffolding, no bones, no clearly demarcated organs. What it did [contain] were metamorphoses, movement, urges, and stagnating resistance" (1991, 124). The medieval body is a fluid body.

Dissection exposes the bodily substance in various stages of metamorphosis: ramifying, differentiating, liquifying, dissolving, putrifying, calcifying. Conducted before its audience of philosophers, physicians, anatomists, artists, aficionados, and onlookers as a performance, a sideshow, a carnival turn, dissection displays the inherence of meaning in the substance of the body. The medieval practice of dissection as spectacle is an extension of the medieval body as spectacular.

The Cartesian alienation of self from body also splits off body from cosmos. The material body and the ethereal self become different orders of event: the body becomes pure substance and the self the absence of substance, insubstantiality, pure absence. The admixture of materiality and ethereality in the medieval body is uncompounded and redistributed into incommensurable substances. Body is no longer cosmos. Disease, separated in the same gesture from its elemental cosmology, is no longer a physical substance but a metaphysical idea with a locus in what Michel Foucault calls classificatory space (1975, 9). Symptoms are clues to a reality which subsists elsewhere. The disease process is only disturbed by its passage through the body (Foucault 1975, 8, 14). Medicine, from being spectacular, becomes speculative (Foucault 1975, xii). Diagnosis is detection. "[W]e can invoke the model of *medical semiotics,* or symptomatology," according to Carlo Ginzburg—"the discipline which permits diagnosis, though the disease cannot be directly observed, on the basis of superficial symptoms or signs" (1980, 12). Disease has become an etherealized essence which can be interiorized in the patient so that it is, as Foucault puts it, "concealed within him like a cryptogram" (1975, 59).

With the reinvention of the body as an object, disease is precisely reinscribed on the tissues (Foucault 1975, 127). "The space of the configuration of the disease and the space of the *localization* of the illness in the body have been superimposed" (1975, 3). No pathological essence exists beyond the symptoms; the symptoms become the disease (1975, 91). "By acquiring the status of object, [the body's] particular quality, its impalpable color, its unique transitory form took on weight and solidity" (1975, xiv). Disease is now entirely inscribed on its surfaces. "Medical rationality plunges into the marvelous density of perception, offering the grain of things as the first face of truth, with their colours, their spots, their hardness, their adherence" (1975, xiii). Autopsy, which has the property of rendering the body nothing but surface, of surfacing the interior, of dismantling interiority, makes the grain of things perceptible throughout its layers. As a consequence, medical knowledge is "redistributed in the space in which bodies and eyes meet" (1975, xi). Medicine is now merely specular, a reflection of and upon the visible. Limiting medical knowledge to the visible protects it from and in the same gesture prevents speculation. Modern medicine supposes that the seeable is sayable without remainder and so exhausts knowledge. Practitioners simply learn to say what they see; discoveries are limited to seeing new objects. Thought is displaced as the locus of knowledge, and perception is limited to surface appearance. The gaze becomes the passive absorption of perception, not the active intellection of seeing. Pain now requires a visible inscription on the body for its authentication. In the absence of this mark, pain is transferred from a disposition of substance, or body, to a disposition of essence, or mind. Thus Cartesian dualism gives rise, according to Nancy Scheper-Hughes and Margaret Locke, to "a tendency to categorize and treat human afflictions as if they were either wholly organic or wholly psychological in origin: 'it' is *in* the body, or 'it' is *in* the mind" (1987, 9).

The equation among the knowable, the visible, and the sayable now seems so obvious that taking them apart again is difficult. But arriving at this perspective required what Foucault called an epistemic break, a shift in the paradigm of knowledge. Thereto, knowledge was a domain of ideas; thereafter, knowledge became a domain of perception. Signs, which used to be clues to the past (anamnesis), the present (diagnosis), or the future (prognosis), became symptoms, manifestations, localizations, instances of the disease. Disease collapses into it surface appearances. Interiority, which used to be mystery, concealment, opacity, is split open and revealed as more surface. The logic of the visible surface is transferred to the interior. But "the forms of visibility have changed" (Foucault 1975, 195). Medical looking is not naive but, as Foucault puts it, an "intellectual eye" (1975, 120), one which reads "the deep structures of visibility in which field and gaze are bound together by *codes of knowledge*" (1975, 90). Duden writes,

> This isolated, objectified, material body was seized by a dissecting gaze that embraced the entire body, not only its surfaces, but also its recesses and orifices. It penetrated inquisitively into the inside, evaluating the palpated organs and relating them to a visual image of the organs of

> cadavers. This gaze turned the body, and with it the patient who possessed it, into a new kind of discrete object. (1990, 4)

Autopsy is the attempt to scrutinize inscriptions on the flesh.

THE CORPSE AS STILL LIFE

Hence, the corpse as still life. When my mother first introduced me to still lifes, I thought they were paintings of things that just happened to be still, cut flowers, bowls of fruit, objects on a table. It did not occur to me till years later, introducing my daughter to still lifes, that the things painted were themselves forms of life but severed at the root, lives stilled, what we have come to call small murders. (The French term, *nature morte,* perhaps captures this possibility more clearly.) Just so with the corpse as with the pear, the wedge of cheese, the haunch of venison, the hung pheasant, the smoked fish: the subjects are arrested in the course of transmuting. They have not altogether lost their character as life forms but have begun nonetheless to take on the character of things, substances, material objects.

THE MIND/BODY PROBLEM

Susan Ritchie (1988) locates the dead body as a site of slippage in our ideology of persons and hence an interesting place to look at subjectivity. Drew Leder writes:

> Whereas my corpse is wrapped in an ineluctable absence, a withdrawal from my self-experience, the corpse of the Other thrusts itself into my view. There is a dys-appearance that renders all perspicuous. The Other's viscera, hidden within bodily depths during life, can now be opened up and explored. Depth disappearance is all but eradicated. So too is surface disappearance overcome. That is, ordinarily I do not thematize the Other's body per se; we exist in a cosubjectivity, directed outward to a common world. But death is precisely what severs (or renders problematic) this cosubjectivity, inaugurating the objectified relation. (1990, 145–146).

The corpse of the Other is given as static, spatially extended, specular. The body becomes the terrain of a disease which it surfaces, literalizes, materializes. The move to traverse that terrain, ambiguously alluded to by its practitioners not as an autopsy but as a pathological examination, is a move to preserve its objecthood by reiterating its ontology as superficial, visible, phenomenal, that which does not hide anything else, which is not transparent to anything else, which cannot be opened to anything else. In the course of this traversal of its anatomy, the corpse nonetheless intimates its subjectivity.

A corpse arrives wrapped and tied. When the technicians unwrap it, her face and genitals have been draped in a white cloth. Her toenails have been recently manicured and painted blood red. An identification tag is tied round her big toe.

The pathologist pats the corpse of an old man on his bald rosy head and then dissects out his brain. After the dissection, the technician stuffs the hollowed skull with crushed paper toweling, remounts the skull cap, unfolds the scalp and stitches it back together. The abdominal cavity is similarly split, stuffed, and stitched. The body rests on slats mounted over a corrugated metal surface like a washboard. Tiny faucets above the head pour water underneath to a drain below the feet. The technician hoses off the body, spraying the face lightly and sponging it down. Water squirts out of the ears. He dries off the corpse with paper towels from head to toe, lays a white cloth over the face, ties the wrists, ankles and chin with strips of cloth and rewraps the corpse in its huge paper sheet, tying the bundle round the neck, middle, and ankles. The shrouded body is lifted onto a cart and rolled back into the refrigerator.

While the husk of a body is thus being reassembled, one pathologist runs loops of intestine through his fingers at the sink. Another is examining the brain on the autopsy table. The technician has the heart out. Two pathologists inspect lungs and kidneys on another table. Viscera, severed, split, sectioned, and sliced, are scrutinized, mounted on slides, slipped into test tubes, dropped into jars. Dr. Wood comes in, looks at the lung on the table and says, "Does she have a pneumonia there?" They speak of her condition, her disease, her death, as if she were impersoned in these slivers of tissue. This is so even after the container for these body parts, the corpse itself, has been closed up and taken away.

After the pathologists finish with the brain, the technician ties one end of a piece of string to it, suspends it in a jar of clear fluid, and winds the other end round the jar's neck, tucking in the loose end of string to which her name tag is attached.

These irruptions of subjectivity through the discourse of objectivity attest to the fragility of the framing of the body-as-object. The toe, that particular gnarled and wrinkled toe with its modest indulgence of feminine vanity, is a locus of personhood. This allusion of the corpse to its own personhood, this insistence on its presence, is attested to by the curious gentleness of the pathologist's pat on the head of the old man, administered with the kind of cheerful absent-minded tenderness with which one pats a child, without any particular expectation of response. The tenderness is reiterated in the way the technician washes the corpse's face, not squirting it directly but spraying it and sponging it gently off. The pang of unease we experience when, as a result of these ministrations, water, sprayed in through the mouth or the nose, comes out the ears of the corpse, likewise signals its not-quite-objecthood. The drapes over the face and genitals, shielding the corpse's most public and particular as well as its most private and generic parts, also describes personhood on the body. The pathologists find personhood in the woman's parts even discorporated, dissevered, and dispersed about the room. Thus the brain, incongruously name-tagged like the toe, bears in its jar an eerie presentiment of the person. Attempts to reach through the frame to the person, or to break out of the frame by the person, trouble the distinction we are culturally concerned to maintain between subject and object, self and body, life and death.

Death, which suggests itself as the moment in time and the locus in space in which the body-as-object is definitively broken off from the body-as-self, is in reality spatially and temporally ramifying. Bodily tissues are affected differentially: vital processes continue to transpire in some tissues while at the same time others become inert and still others begin to undergo processes of decay. Foucault remarks, "Death is therefore multiple, and dispersed in time: it is not that absolute, privileged point at which time stops" (1975, 142). It encroaches on us. It exscribes the edges of our universe of discourse. The corpse portends a presence which is also an absence. Hence Julia Kristeva's claim for the corpse as *abject.*

> In that compelling, raw, insolent thing in the morgue's full sunlight, . . . I behold the breaking down of a world that has erased its borders: fainting away [from *cadere,* to fall, and so akin, according to Kristeva, to cadaver (1982, 3)]. The corpse . . . is the utmost of abjection. It is death infecting life. Abject. It is something rejected from which one does not part, from which one does not protect oneself as from an object. Imaginary uncanniness and real threat, it beckons us and ends up engulfing us. (1982, 4)[7]

It is this ambiguity with which we feel uncomfortable. Pathology's attempt to reinscribe the space of death on the body as precise, to rearticulate the time of death as instantaneous, is a response to this discomfort. These are attempts to disambiguate the ontological status of the corpse as subject-turned-object by conjuring up crisp, clear, clean boundaries in space and time. We would be more comfortable with an unambiguous shift, like a magical transformation into elemental particles, ash for instance, which accounts, perhaps, for the popularity of cremation. The tenuousness of the reinscription dictates the concealment of the death, the corpse, the autopsy. Fear of death is fear of ambiguity. The corpse, incorrigibly carnivalesque, arouses this fear.

DEATH AND THE CARNIVALESQUE

When the chief pathologist escorted me through the underworld to the morgue, he warned me against giving way to unseemly merriment, cracking jokes, mocking the corpse. Nothing was further from my mind. The autopsy seemed to me entirely unfunny, awesome if anything, nor was I tempted at any time in the course of it to make a joke. However I realized, on reflection, that if, under the tension of suppressing the shock of my first encounter with a corpse, someone else had offered up a bit of black humor, I might well have burst out laughing.

The technician, washing down the table after an autopsy, said, "As far as I'm concerned, once the," he hesitates, "soul leaves the body, that's it." The view participants in the discourse of pathology put forward is that nobody is there, just an arrangement of tissues. Yet this reading is threatened by the way the body makes allusion to the person: the mystery of embodiment. The nerve end of this allusion, this threat, appears in gallows humor. Richard Zaner writes, "with the corpse we face a 'sensed relic' of the once-embodied and communicating person—whence, of course, the otherwise quite unaccountable gallows humor" (1981, 30). Gallows humor comes off only under the eerie assumption that somebody is still there to be insulted, profaned, abused, mocked. Insults can only be offered the self, not the object. Evidence of this metaphysical unease about the ontological status of the corpse appears in the prohibition against black humor.

Death in the carnivalesque is approached in terms of "food, drink, ritual (cultic) indecency, ritual parody and laughter" (Bakhtin 1981,218–219), which forms the basis of comedy. Death becomes a "gay monster" (Bakhtin 1984, 151n) in which fear of death is defeated by laughter (Bakhtin 1984, 91). This laughter, this assault on the sacred, the aristocratic, the pure discourse, prevents it from becoming authoritative. "Laughter purifies from dogmatism, . . . the single meaning, the single level, from sentimentality" (Bakhtin 1984, 123). Black humor threatens not so much the dignity of the corpse as the aristocracy of the discourse that inscribes it, the discourse of medicine. The threat of the carnivalesque must be vanquished by the discourse of medicine in order to ensure its aristocracy. Medical apprenticeships entail learning to overwrite the corpse with the new discourse. The pathologist describes the evisceration of the old man in the autopsy report as follows.

> The usual "Y" shaped incision is made disclosing musculature with increased hydration and a panniculus measuring 4 cm. The organs of the thoracic and abdominal cavity are in their normal anatomical position. There is a large amount of straw colored fluid in the chest cavity bilaterally. There are adhesions noted in the apex of the left lung and a 150 cc. of fluid are noted in that thoracic cavity. On the right side there are adhesions all over the lateral surface of the right lung which are fairly dense and the 300 cc. of straw colored fluid is noted in that side. There is dependent stasis on both lungs. (Autopsy Report, 1987)

And with respect to his brain autopsy:

> The scalp is reflected in the usual manner and the bony cover appears intact. The pachymeninges are thin, fibrous and glistening. There is no evidence of previous laceration, contusion or healed fracture. The cere-

> bral hemispheres are rounded and reveal normal convolutions and sulci. The venous tributaries are markedly congested. The brain weighs 1360 gm. The brain parenchyma is soft, smooth and glistening with no evidence of hemorrhage or petechiae. The vessels at the base of the brain are unremarkable. (Autopsy Report, 1987)

The brain of the African-American woman is described thus:

> Both external and internal surfaces of the dural leaflets are patent. There is no evidence of herniation of the cingulate gyri, unci or cerebellar tonsils. The leptomeninges are thin, translucent and free from exudate. Examination of the arteries of the circle of Willis and their major branches reveals they are patent and free from atherosclerosis. The superficial veins of the brain and the cranial nerves are unremarkable. (Autopsy Report, 1987)

For initiates, the carnivalesque body is not there, only its anatomical reinscription as the proper body. Concealment of the corpse is a recourse against the uninitiated who might catch its carnivalesque qualities. The discourse's very establishment of itself as aristocratic sets it up for assault by the carnivalesque, an assault which translates the high, refined and pure as pompous, stuffy and snobbish. Like all aristocratic discourses, medicine is dependent on what it excludes (Hegel 1977, 111–119; Stallybrass and White 1986, 5). The low, the profane, the grotesque stand as the discourse against which it defines itself. At the same time, they constitute the discourse to which it is vulnerable. Autopsy is threatened by the breakthrough of grotesque realism, by the thought, the hint, the whiff of sausages.

CONCLUSION

• • •

THE MAKING OF THE MEDICAL BODY

. . . mindless bodies on the one hand and disembodied minds on the other—are unnatural species.

—*Maxine Sheets-Johnstone*

Our epistemology of the body is Cartesian. We tend to think of the body as a sort of solid object and of the self, and in particular, the mind, as a sort of ethereal subject mysteriously infused into the object. The result of this curious dismantling of person into subject and object is called, in philosophy, the mind/body problem. If the mind is rarified, insubstantial, and ethereal, and the body is solid, substantial, and material, how do we hook the mind onto the body, get the essence into the substance, put presence in the flesh? What is the relation between the mind and the body?

Descartes put forward self-reflection as the constitutive experience of personhood: "cogito ergo sum" ([1641] 1962, 28). In so doing, he set aside the constitutiveness of bodily experience. This foundational gesture privileges moments of reflection, meditation, and abstraction, and discounts moments that thematize the body in our awareness. Sensory modalities, Descartes holds, take impressions of objects in the form of perceptions. These are not properly thoughts in the Cartesian sense of innate ideas; they are what Descartes calls adventitious ideas, brought to mind by imagination in the form of images, rather than by intellection in the form of reasoning ([1651] 1962, 44, 65). Thought, on Descartes' argument, is independent of the body; perception, by contrast, is contingent on the body's impressionability to objects. The exclusion of sensory modalities from the act of reflection produces the tautology that disembodied thought is constitutive of personhood. The trick, then, is to resituate thought in the body.

As the site of intersection in the body for the mind, Descartes proposed the pineal gland, a proposition whose absurdity seems to us patent. Yet we have transferred this site, with its still unexpunged absurdity, to the brain. In consequence, we have come to misconceive the brain, as if it were different in kind from other perceptual organs. Resisting this disposition, Drew Leder describes the brain is a sensory receptor, but one in an extreme state of reticence. "As with any other corporeal structure—the eye, the mouth, the hand—the brain can be understood not only physicalistically but as an organ of the lived body, a structure of possibility opening onto the world" (1990, 111). It is the most profoundly recessed site, not only of our mullings and conjurings of the world but also of those sensory perceptions which surface at the eyes, the ears, or the skin. Like these other perceptual loci, the brain disappears in moments of sentience; unlike these others, it cannot be recalled to our awareness. "The brain," in this respect, "shares with the viscera a depth disappearance" (Leder 1990, 113). It is perhaps the most incorrigibly absent organ of the body. "That organ which subtends embodied consciousness is itself unavailable to conscious apprehension. Yet this disappearance is nothing but the radicalization of a phenomenon already discussed: the focal disappearance of the surface body. That from which I perceive and act necessarily resists my thematization. As I gaze upon the world, I cannot see my own eyes" (1990, 113). Thought, in the sense of introspection, and perception, in the sense of awareness of external objects, are both modes of absence from the body, and specifically, from the brain.

Leder argues, "This invisibility of the brain is one experiential source for the notion of the human mind as immaterial" (1990, 115). Mind/body dualism polarizes phenomena that are only modestly differentiated in experience. "Times in which the body is most tacit and self-transcending are collected under the rubric of rational 'mind.' Other experiences, where corporeality comes into strong thematic presence, are collected under the rubric of 'body' " (1990, 149). Both aspects of the mind/body problem, the etherealization of the mind and the materialization of the body, have their roots in our experience of embodiment, as the body absents itself from or obtrudes itself on our awareness.

Descartes holds that the reflective faculty is neither the imprint of the world on the mind nor even the elaboration of the world out of the mind, but, mysteriously, the absence of any imprint. Against Descartes' claim that thought transpires prior to bodily experience, set Maxine Sheets-Johnstone's contention that thought is rooted in bodily experience. "Concepts were either generated or awakened by the living body in the course of everyday actions such as chewing, urinating, striding, standing, breathing, and so on" (1990, 4). These tactile-kinesthetic experiences of the body give rise to what Mark Johnson describes as "imaginative structures of understanding, such as schemata, metaphor, metonymy, and mental imagery" (1987, xi). They do so by a method akin to the one Victor Turner describes for ritual symbols (1970, 48–58).

Symbols are polarized into two clusters of properties, a sensory cluster and an ideological cluster. The sensory properties are grossly physiological or natural qualities, closely related to the form of the symbol, which arouse desires or feelings. Turner calls these concrete or orectic properties, from the Greek *orektikos,* meaning appetitive, of or pertaining to the desires. The ideological properties consist of moral norms or principles of social structure, structural relations, which order thoughts. Turner describes these as abstract or normative properties. By virtue of the shift from the concrete to the abstract pole, Johnson writes, "we make use of patterns that obtain in our physical experience to organize our more abstract understanding" (1987, xv). Thus "what is typically regarded as the 'bodily' works its way up into the 'conceptual' and the 'rational' by means of imagination" (1987, xi).

The terms of rationality are at root corporeal, preeminently visual, as in "I see what your mean," but also auditory, as in "It sounds all right," gustatory, as in "I can't digest it," or postural, as in "I under-stand it." On present reckoning, these are not quite metaphorical remarks. Metaphors proper propose a relation between two disparate realms; these corporeal metaphors propose that the realms are not disparate. The instance of embodiment shifts us from the suggestion that thought is like looking, listening, eating, or standing to the notion that thought partakes of these.[1] The terms of the metaphor are not polarized but are articulated along a spectrum from concrete to abstract as bodily moves toward intelligibility.

Not only is thought rooted in bodily experience but also bodily experience is rooted in thought. As an idealist, Descartes privileges mentality and then reconstitutes the realm of objects in the mind, granting that realm contingent status. Clarity and distinctness are the attributes of an idea, not of an object, and attest to the existence of mentality, not materiality:

> [A]lthough I certainly do possess a body with which I am very closely conjoined; nevertheless, because, on the one hand, I have a clear and distinct idea of myself, in so far as I am only a thinking and unextended thing, and as, on the other hand, I possess a distinct idea of body, in as far as it is only an extended and unthinking thing, it is certain that I (that is, my mind, by which I am what I am) is entirely and truly distinct from my body, and may exist without it. ([1641] 1962, 69)

Materialists have reversed Descartes' privilege, granting the realm of objects priority. The business of the mind, on that view, is simply to apprehend objects. But, as Jonathan Miller points out, perception does not merely passively take the impression of already clear and distinct objects; thought renders objects clear and distinct. Perception is a process of continuous self-correction. What actually presents itself to the eye, for instance, is hazy, unsteady, and discontinuous.

> The projection area, then, is not simply a camera obscura, carrying a miniature replica of what it represents: it is a mathematical analyser which breaks down the physical components of the sensory input into the biologically relevant features of the physical world. Comparable analyses occur in each of the primary projection areas, visual, auditory and tactile.
>
> Such analyses can work only on the information which the sense organs themselves provide—spatial patterns of illumination in the case of the eye and temporal sequences of vibration in the case of the ear. And yet the perceptual experiences of vision, hearing and touch are incomparably richer than that, which suggests that perception is something over and above the mere inspection of all the information which is displayed and analysed in the primary projection areas. (Miller 1978, 335)

We fill the spaces between perceptible partials, adjusting and readjusting the pattern until it is clear and distinct. It is not the objects that are clear and distinct but, as Descartes insisted, the ideas. Perceptual modalities are conjectural as well as sensory. The brain "reproduces not patterns but fictional objects" (Miller 1978, 338). So the body is not divorced from thought but continuous with it, sometimes obtrusive and sometimes recessive in our experience of thinking. The Cartesian cleft between thought and the body dismantles our access to the corporeal "roots of thinking," in Sheets-Johnstone's phrase (1990). Insofar as invention participates in perception and perception in thought, however, this split is unwarranted even in Cartesian terms.

What we have taken from Descartes is not his own struggle with the mind/body problem, his attempt to make sense of their intimate conjunction, expressed in his image of the pilot in his ship, but only the opposition between two substances held to be different in kind: *res extensa,* extended things, which we take to be objects, one of which is the body; and *res cogitans,* thinking things, which we take to be mind, self, or thought.

> For when I think that a stone is a substance, or a thing capable of existing of itself, and that I am likewise a substance, although I conceive that I am a thinking and un-extended thing, and that the stone, on the contrary, is extended and unconscious, there being thus the greatest diversity between the two concepts,—yet these two ideas seem to have this in common that they both represent substances. ([1641] 1962, 49)

Of course substantiality turns out to be just what they do not have in common. What Descartes means by *res extensa* is not objects but the idea of objects, which he calls "corporeal nature." And what he means by *res cogitans* is not substance but perfect insubstantiality, absence. Both are abstractions which remain recalcitrantly mental in contrast to objects which remain obdurately physical.

But the difficulty for Descartes in starting from either mind or body is to arrive at the other. If Descartes thinks of his body as "a countenance, hands, arms, and all the fabric of members that appears in a corpse" ([1621] 1962, 38) and then he thinks of this body as animated by a soul, which he imagines as "something extremely rare and subtile, like wind, or flame, or other, spread through my grosser parts" ([1641] 1962, 38), then he is perturbed by the mind/body problem in the form of the incommensurability of substances: how is the "rare and subtile" substance spread through the grosser one? He assuages this perturbation by granting priority to the mind and considering that its perceptual faculty attests indirectly to the existence of the body. Importunings, not of things but of the body, its sensations and perceptions, evidence his body to his mind. "Nature teaches me by these sensations of pain, hunger, thirst, etc., that I am not only lodged in my body as a pilot in a vessel, but that I am besides so intimately conjoined, and as it were intermixed with it, that my mind and body compose a certain unity" ([1641] 1962, 70). It is this unity, and also this disparity, that we have yet to address.

We have become attached to an interpretation of Descartes to which Descartes himself would not have subscribed, namely, the notion that the body is an object. Having wrestled with the incommensurability between a purely material body and a perfectly insubstantial mind, Descartes concluded that the body is, in its primary presentation, an idea in the mind. The materiality of the body, along with the other objects that are impressed upon the bodily substance, is, in effect, a literalization of this idea in our experience.

The body, we might say, just comes to mind. As soon as we turn our attention to it, we become aware of the body as an object. When we attend either to things or to thoughts, the body slips away, in effect, dematerializes. We simply cannot turn our attention *to* the body as subject; as subject, the body is what we turn our attention *from.* In medicine, the body can be materialized and dispossessed of mind, as in medical examinations, surgical procedures, or pathological examinations, and more rarely, the self can be etherealized and lifted out of the body, as in gynecological examinations or narratives of the self. These moments of either materialization or disembodiment are elaborate, fragile, and transient constructions. Neither is the ordinary realm-status of the body. The mindless body is evoked when the body is thematized in consciousness just as the disembodied mind is evoked when the body disappears from awareness. Thematization and disappearance are not ontological oddities, suspect states, but merely two of a spectrum of possibilities of embodiment.

Suppose we start elsewhere, neither from the reality of objects nor from the primacy of ideas, but from embodiment. From there, not only does the body never reduce itself to the paradigm of physical object, but also it resists etherealization into an idea. "Body" is clearly the perverse term in this discourse. I am not enveloped in materiality but materialized corporeally. Apparent oxymorons like "psychosoma," "lived body," "embodied self," or Nancy Scheper-Hughes and Margaret Locke's elegant term "mindful body," are, as they write, attempts to express "the myriad ways in which the mind speaks through the body, and the ways in which society is inscribed on the expectant canvas of human flesh" (1987, 10). From this perspective, the philosophical problem is not forcing together the mind and the body; it is taking them apart.

And body is the pivotal term here. Juxtaposed to "mind," "body" takes on materiality; juxtaposed to "object," "body" takes on subjectivity. The mind/body problem invents itself in its own terminology. The proper contrast is not mind/body but body-as-self/body-as-object. Materialization, like its occasional corollary, disembodiment, are both moves away from the embodied self. The body-as-object is only perverse when it is put forward as the natural order of the body, rather than as the specialized inflection imparted to it by, among other discourses, medicine. Commonly, in medicine, as in everyday life, when we become aware of our own materiality, what we are aware of is not the body as an object but the corporeality of the self. We are precipitated into our own skins, neatly embodied, acutely present in our own flesh.

CODA

• • •

PERSPECTIVES ON EMBODIMENT

To write the body.
Neither the skin, nor the muscles, nor the bones,
nor the nerves, but the rest: an awkward, fibrous,
shaggy raveled thing, a clown's coat

—Roland Barthes

The root puzzle in ethnographic writing is how to get access to the Other. But the Other is a category invented by social scientists to create an opposition between us and our objects of analysis. They become inhabitants of another universe of discourse, one we not only observe but also fabricate. Rendered marginal, eccentric, strange, the Other constitutes the "outskirts of the familiar" in terms of which we define our own centrality (Natanson, 1970, 37). The category of the Other is conjured up to provide what Amy Shuman has called "the place for the unknown" (1990, pers. comm.). It is this invention of the category of the Other that makes access the root puzzle.

This marginalization of the Other permits the Bakhtinian gesture by which the marginal is recast as the low. We do not so much look *at* as look *down on* the Other. A hierarchy of discourses is instantiated. Sacred or aristocratic discourse is raised symbolically: purified, etherealized, and disembodied. This high discourse becomes the locus of privilege, power, and authority.

> The authoritative word demands that we acknowledge it, that we make it our own; it binds us, quite independent of any power it might have to

> persuade us internally; we encounter it with its authority already fused to it. The authoritative word is located in a distanced zone, organically connected with a past that is felt to be hierarchically higher. It is, so to speak, the word of the fathers. Its authority was already *acknowledged* in the past. It is a *prior* discourse. (Bakhtin 1985, 342)

By contrast, profane or vulgar discourse is symbolically lowered: degraded, materialized, corporealized. The low discourse becomes a source of ambiguity, contamination and outrage, a discourse Mikahil Bakhtin calls the grotesque.

> The essential principle of grotesque realism is degradation, that is, the lowering of all that is high, spiritual, ideal, abstract; it is a transfer to the material level, to the sphere of earth and body in their indissoluble unity. (Bakhtin 1984, 21–22)

The relation between these discourses is akin to Hegel's master/slave paradox (1967, 34–240). On the one hand, the master depends on the slave for the position of mastery. That position is wholly contingent on the existence of the slave. On the other hand, the master holds the slave in the position of slavery by force, foisting on the slave a position the slave neither conceives nor chooses. Hence the high discourse is at once dependent on the low and threatened by insurrection from the low.

The high discourse is a colonial discourse, designed to suppress, subjugate, and overcome the Other. It is a discourse of conquest. When a conquered population is incorporated into the social body of the conqueror, the residue of colonialism can survive not only in the social hierarchy but also in a linguistic one. The status of Latinate words over Anglo-Saxon in English, for instance, is presumably an aftermath of the Norman Conquest. In cultivating a Latinate vocabulary, scholarly language partakes of the authoritativeness of the conquerors.

The Other originally constructed by colonial discourse is the savage, the primitive, the native.[1] The terms transmute with the lineage of explorers, missionaries, anthropologists, and tourists (MacCannell 1990, 18). By virtue of this construction, we ourselves are constituted the civilized, the modern, or the scholar. These constructions make the relationship between the discourses of self and Other hierarchical and estranged.

Where antithetical discourses are juxtaposed, the hierarchy can be disturbed. The disturbance consists of an assault by the low on the high, resulting in inversion, reversal, overthrowing, undermining, transgression. The high is brought down, overturned, mocked, abused, inmixed with the low. What Bakhtin calls a carnivalization of discourses occurs (1984, 17). The boundaries between discourses become blurred, fragments migrate across domains, through hierarchies. Words bear traces, carnivalesque encrustations, glimmers of their history (1985, 276).

> Indeed, any concrete discourse . . . is entangled, shot through with shared thoughts, points of view, alien value judgments and accents. The word, directed toward its object, enters a dialogically agitated and tension-filled environment of alien words, value judgments and accents, weaves in and out of complex interrelationships, merges with some, recoils from others, intersects with yet a third group: and all of this may crucially shape discourse, may leave a trace in all its semantic layers, may complicate its expression and influence its entire stylistic profile (Bakhtin 1985, 276)

There is at once a tension and a contagion among words, between words and their objects, and between words and their interpreters. "A word forms a concept of its own object in a dialogic way" (Bakhtin 1985, 279). It takes, as David Richter puts it, the imprint of the Other (1986, 412).

> As a living, socio-ideological concrete thing, as heteroglot opinion, language, for the individual consciousness, lies on the border between oneself and the other. The word in language is half some else's (Bakhtin 1985, 293)

The effort of an authoritative discourse is not only to repudiate the carnivalesque, to strip off the encrustations of words in the interest of a pure discourse, but at the same time to preserve the carnivalesque elsewhere as a lodgement for the category of the Other.

Ethnographic writing is explicitly constructed in order to get access to the Other. There are two difficulties with this construction. One is the estrangement of self from Other, ethnographer from native. The problem of access to the Other is in part an artifact of the invention of the category of the Other. The other difficulty is the resistance of the discourse to this estrangement. Despite its move in academics toward univocality, homoglossia, and monologism, discourse remains obdurately dialogic (Bakhtin 1985, 282). Peter Stallybrass and Allon White investigate "hierarchies of high and low in the human body, psychic forms, geographical space and the social formation" (1986, 2). These discourses are entangled with each other. The perverse dialogism of words is reflected in the scandalous carnivalization of the body. What dialogism finally challenges is the category of the Other.

The dialogue in question here is between the ethnographer and other personae of the ethnographic text. These persons are constructed narratively, either by disclosing a trajectory that takes a narrative shape or by being enclosed in a trajectory given a narrative shape by the ethnographer. The resulting "self-narratives," to use Kenneth Gergen and Mary Gergen's term (1983, 255), subsume the identities of persons under the aesthetic conventions of narrative (Gergen and Gergen 1986, 3). These narrative selves take shape within forms of life the Gergens also characterize narratively as tragedy, comedy, or romance (1986, 8) or out of a more fundamental disposition Jerome Bruner conceives of as the "narrative mode of thought" (1986, 28). Listening to people tell stories about their lives, Bruner and his colleagues noted, "we discovered that we were listening to people in the act of *constructing* a longitudinal version of Self" (1990, 120). This supposed narrativity of the self warrants the turn in ethnographic writing toward the use of narrative devices to describe the Other. After all, the Gergens argue, "the fact that people believe they possess identities fundamentally depends on their capacity to relate fragmentary occurrences across temporal boundaries" (1983, 255). In adopting the devices of narrativity, the ethnographic enterprise takes its legitimation from the narrative habits of everyday life.

But the self is not a story, or not just a story. A person's life may have an identity, and an identity of a narrative kind, but that identity is not the same as the person who has it. Richard Wollheim points out:

> Indeed, many philosophers have been so preoccupied that they haven't always noticed whether they were talking about a person and his identity or about a person's life and its identity. They reveal this when they take what they have convinced themselves is a perfectly satisfactory unity-relation for a person's life and re-employ it, without adjustment, as the criterion of identity for a person, and thus finish up with a view of a person as a collection of events spread over time, which cannot be right. (1984, 11)

Persons occur at a nexus of spatial as well as temporal phenomena. They cannot be reduced either to their status as objects extended in space or to their status as narratives extended in time. A person might express a narrative sense of self but the person still inheres at a point of intersection between that narration and the body. As Wollheim puts it, "living is an embodied mental process" (1984, 33). Bodies can be invoked as both source and site of discourses of the self. Stories are corporeal acts; the body gives rise to narratives. And bodies are themselves narrated, discursive, inscribed; stories give rise to the body. "Interrelatedness of event is supplemented by identity of person" (Wollheim 1984, 19).

Narrative presentations of self differ from narrative representations of the ethnographic Other specifically in respect of the body. Telling stories about one's self expands aspects of a person already present in the flesh. The teller, as Erving Goffman describes it, exudes a second self (1974, 511–537) but one that alludes to as well as distinguishes itself from its presenter. Writing about the Other, by contrast, contracts the presentation of a person to its narratable aspect. Narrative presentations of self might be held to evidence an impulse toward narrative thinking. Narrative representations of someone else, though, may merely impute narrative conventions to descriptions of persons. Narratives of the self taken out of their context of performance and narratives of the Other by social scientists suffer the same hiatus: an absence of the body. Consider, then, how ethnographic narratives conjure up persons, make bodies, flesh out a self, or body forth the mind of the Other.

Ethnographic writing characteristically presents discursive passages, in which the ethnographer establishes authority in virtue of command over the material, interspersed with scenes, in which the ethnographer guarantees authenticity by attesting to having been there in the flesh (Clifford 1983, 118, 128; Marcus and Cushman 1982, 29). These scenes, consisting of elaborations of settings, descriptions of persons, accounts of events, and passages of dialogue, are held to provide the opening in ethnographic writing for the voice, the presence, of the Other. Here the ethnographer is exscribed, the Other bodied forth.

The claim that such scenes constitute transparencies to another realm rides on the Aristotelian assumption that language imitates reality. But the sense in which language, or any of its parts or arrangements, has parity with reality, or any of its parts or arrangements, is, at the least, mysterious

(Rimmon-Kenan 1983, 106–107). But there is a possible exception: dialogue. Arguably, written dialogue does mimic spoken dialogue. In it, discursive time elapses with real time so that something on the order of a one-to-one correspondence between text and world can be contrived. Indeed, on the Platonic argument, only dialogue is mimetic. There, the voice of the Other might become audible.

On the theory that scenes represent at least some aspects of realities, ethnographers set into their discursive writings occasional narrative evocations of the realm of the Other. Multiple realities are thereby fitted together like Chinese boxes: writings in a realm of discourse are held to be transparencies to a realm of events for readers in a scholarly realm (see Marcus and Cushman 1982, 51). Thus Jean-François Lyotard writes, "an essentially realist epistemology, which conceives of representation as the reproduction, for a subjectivity, of an objectivity that lies outside it—projects a mirror theory of knowledge and art, whose fundamental evaluative categories are those of adequacy, accuracy, and truth itself" (1984, viii). Such scenes can then serve as keystones for an architecture of realistic description constructed around them (Marcus and Cushman 1982, 40–41). In describing the realm of medicine, the scene of a single medical examination might be played out (but see Marcus and Cushman 1982, 35).

The realm of the Other, however, is not constitutive of ethnographic writing but constituted by it. In the act of writing, I conjure up an alternate reality, one I then display to you as reader. The point of access to both the realm of discourse and the realm of events is writing: relationships between ethnographer and native are textualized as characters in a scene; relationships between writer and reader are textualized as narrator and narratee. A Goffmanesque twist turns up in the relationship between these realms: I-as-writer conjure up a realm which is inhabited by myself-as-character (1974, 512). But the imponderables of the ethnographic enterprise appear primarily in the relationship between the self-as-writer and the Other-as-character and secondarily between the self-as-writer and the Other-as-reader. The power of writer over character raises issues of biases, impressions, distortions, misrepresentations, elisions and elaborations, mistakes and inventions: issues, in respect of the relationship between ethnographer and native, of subjectivity; issues, in respect of the relationship between writer and reader, of fictionalization. Realistic writing is cunningly constructed to deflect imputations of fictionalization. Yet realistic writers appear naive about the conventions they use to sustain an impression of objectivity.

Consider, then, the uses of narrativity in three scenes, each the examination of a woman's breasts by a male physician. The scenes are taken from transcriptions of audiotape and/or descriptions of observations of medical examinations. They bring together two discourses, a discourse of the self and a discourse of the Other, variously transmuted, interrupted, contaminated and encrusted by each other and by other discursive configurations. All claim access to the Other but each from a different angle.

SCENE 1

Realm:	Medicine.
Scene:	The medical examination.
Setting:	The doctor's waiting room, connected by a hall to his office and his examining room.
Characters:	Dr. Silverberg, the internist; Mrs. Gillette, the patient.
	((It is late afternoon. Mrs. Gillette sits alone in the waiting room. She is a tiny, slight old woman with a long equine face, crisp grey hair, and gnarled hands, the fingers bent at the tips. There has been some sort of a mix-up in the scheduling and she has been waiting all afternoon to see the doctor. As the scene opens, Dr. Silverberg comes out into the waiting room.))
Dr. S:	Last but not least.
	((With a slight bow, he sweeps open an arm toward the hall to his office. He is a thin middle-aged man with a narrow, dark head and fine features, closely cropped hair, glasses, and a serious look. He wears a white coat over a three-piece suit. On the way down the hall, they talk about Mrs. Gillette's son, Fred, whom the doctor knows. In the office, Dr. Silverberg settles Mrs. Gillette in an armchair across from his desk.))
Dr. S:	How old are you.
Mrs. G:	Ninety—
	Oh seventy.
	((She giggles. A slight air of disorientation attends her remarks.))

Dr. S: What can I do for you.
Mrs. G: My— (huh)
My doctor—
I have a doctor comes to my house every month (   )
Dr. S: Yeah.
Mrs. G: Fred thinks he don't come for anything.
Dr. S: O.K.
Mrs. G: And
a windy day this was— a real windy day but I—
got severe pains in this chest.
((She describes her pains. As she speaks, Mrs. Gillette leans across the desk toward the doctor, resting one arm along the arm of her chair and gesturing with the other, touching her body as she speaks of it. She has a fluttering birdlike way of moving and a quick episodic style of speech. Dr. Silverberg finishes taking her history and then escorts her down the hall to his examining room. This contains a high narrow examination table across from a medicine cabinet. Next to the cabinet is a chair and at the head and foot of the examination table, two stools. The doctor directs Mrs. Gillette to the chair and hands her a folded paper gown from a drawer under the examination table.))
Dr. S: And I'd like you to take
everything off
and put this gown on so that it opens in the back.
All right?
Mrs. G: Umhm.
Dr. S: And I'll be back in just a minute.
((He leaves, closing the door behind him. When he comes back, he knocks and opens the door. Mrs. Gillette jumps lightly up onto the examination table, carefully holding her gown closed at the back. Dr. Silverberg comes over to her and picks up her hands, turning them to inspect them. He takes out an otoscope to check her ears, holding her head with one hand and the instrument with the other. Then he dismantles the otoscope and mounts an ophthalmoscope on the same stem, touches the top of her head

with his left hand and directs the instrument in turn at her left and right eyes, resting his right wrist on her left shoulder. Putting the instrument away, he shifts his hold to the back of her head and palpates the glands in the front of her neck with the other hand. Then he perches up on the table behind her to listen to her heart, his left hand, fully open, holding the stethoscope against her back, his right lightly holding the front of her shoulder. To examine her chest, he steps down, takes hold of the back of the same shoulder with his left hand, and has her lie down.))

Dr. S: Let me help you off with this.

((He draws the gown off her shoulders and down her arms to expose her breasts.))

Mrs. G: (I'm a prude.)

Dr. S: And I'll just examine carefully your breasts.

((As he leans over to examine her left breast, she jerks away, curling up her left shoulder.))

Dr. S: Try to relax= I'll be as gentle as I can O.K.?

Mrs. G: O.K.

((He touches her left breast with the tips of his fingers. With a perceptible effort, she holds her body still, the left shoulder still curled with tension.))

Mrs. G: I fell on there= no wonder.
On the point of the banister.

((He moves his hands over to examine the unbruised right breast.))

Mrs. G: Tripped.

((Now she is quite relaxed. A lump the size of a walnut is visible between the doctor's fingers under the papery skin of her other breast.))

OBJECTIVITY

The conventions employed to construct ethnographic descriptions are at root fictive (Marcus and Cushman 1982, 28–30). I shall investigate two such conventions, perspective and voice, which Shlomith Rimmon-Kenan distinguishes as who sees and who speaks (1984, 72). Perspective is the locus in space from which the perceiver sees the events described and, by extension, the moment in time from which the perceiver unfolds those events. This locus can be external to the realm of events or internal to it and without or within the bodies of characters. Perceptual perspectives imply epistemological, emotional, and ideological ones (Rimmon-Kenan 1984, 71–82).

Perspective

Realistic writing tacks from its defining external anchorage to an unanchored perspective internal to the realm of events but external to the bodies of its characters, as if the perceiver had been lowered into the scene from an aerie above. From the external perspective, the realm of events appears as a microcosm of whose parts the perceiver has a simultaneous or panoramic view and of whose processes the perceiver has a transtemporal or atemporal awareness. The realm is seen from another space and time by a perceiver who has unrestricted access to its spaces and times, access not only to what transpires within the realm, but to what has transpired or will transpire beyond its boundaries. This perspective has been called in literary theory the bird's eye or God's eye view. The shift to an internal perspective moves the perceiving eye into the realm of events but without the bodies of characters and so that the perceiver floats omnipresently around what transpires there. This perspective restricts spatio-temporal perception to the horizons of the taleworld: spaces appear from a locus within the realm and time unfolds over the course of events.

In the description of the medical examination, the external perspective is initially established by the perceiver's knowledge of the topography of waiting room, the doctor's office and examining room, and, by implication, the topography of the hospital, the city, in short, the realm of events. The external temporal perspective is apparent in the perceiver's knowledge of what has transpired in the afternoon before the scene occurs, of what will transpire in the scene before it does, and, presumably, what did transpire afterward. Knowledge is also external. Not only does it include local knowledge of the names, appearances, and statuses of the characters before they appear on the scene but also transcendent knowledge of the realm of medicine and its status in the realm of events. The perceiver's emotional detachment from the scene indicates an external emotional perspective. Whatever flushes of feeling the characters experience do not affect the perceiver. Grasping the scene, as it were, from the outside or above grants the perceiver ideological superiority. The realm of medicine, however authoritative in its own right, is presented here as a realm whose ontological conditions can be transcended by its perceiver.

The scene shifts from an external perspective to an internal perspective from without the bodies of characters as it unfolds. The perceiver floats disembodied about the scene. From here, objects appear up close from various angles: the patient's back, the glands of her neck, the top of her head. Spatial perspective is contained within the boundaries of the scene but is not otherwise restricted. Temporal perspective is characteristically contained by the scene and unfolded over the course of it. Events are represented in the order in which they transpire. Knowledge is acquired as the scene unfolds. Transcendent knowledge, emotional detachment, and ideological superiority are preserved so that the shift to close focus occurs without loss of objectivity. The ethnographer is inserted into the scene but invisible in it. To this extent, the internal perspective from without the bodies of characters can be regarded as an inflection of the external perspective, with a close rather than remote focus.

Voice

Voice is the conceptual locus of the narrator who describes the events. The narrator can speak from outside or inside the realm of events, loci distinguished by Gerard Genette as extradiegetic and intradiegetic, and can participate in the events or not, loci distinguished as homodiegetic or heterodiegetic (1983, 227–237). Narrative voice can be perceptible or imperceptible in the text and reliable or unreliable (Rimmon-Kenan 1984, 94–103). The convention in realistic writing is to have an extradiegetic, heterodiegetic narrator whose voice is imperceptible and reliable. The narrator speaks as an outsider of the inside of another realm, as an observer of events in which the Other participates. Speaking so separates self and Other, alienates, specifically, their voices. The objective voice is arrogated to the narrator; the subjective voice attributed to the Other. This strategy enhances the authority of the narrator at the same time that it enhances the authenticity of the native.

Narrative voice is barely perceptible in this text except in the act of description. It is detectable in assessments of attributes of the scene: the air of disorientation, the quality of the holds, the appearance of tension or relaxation, and the like. Over the course of the scene even this slender narrative voice attenuates as the characters' voices intensify. Imperceptibility in the text enhances the narrator's claim to reliability since no intervening intelligence appears to skew description. The internal perspective and intradiegetic voice are embedded in this scene in the external perspective and extradiegetic voice so that omnipresence and detachment serve as the framework for close description.

From this complex of holds, objectivity emerges as a central concern in ethnographic writing. The self is decentered in order to center the Other. Bodily withdrawal from the realm of events or disembodiment in it are the marks of objectivity. Bodily withdrawal leaves behind the spoor of the voyeur: though the body vanishes, its perceptual apparatus, especially the eyes, remains. Ethnographic writing instantiates an invisible perceiver in a visible world. The ethnographer sees what transpires between physician and patient, sees even the intimacies of the examination, without being seen, without materializing or appearing. By contrast, characters in the scene are staged figures, minutely and globally visible. The scene itself is opened, tilted, pivoted for inspection. In a visually constituted realm, the thrust of writing is to *tell* what the ethnographer has *seen.*

Authority

Perspective and voice are so connected in realistic writing that the external seer speaks in an extradiegetic voice; the extradiegetic speaker sees from an external perspective. The external perspective is understood to lie close to the narrating agent (Rimmon-Kenan 1984, 75). This positioning creates a tacit textual claim to transcendent authority. The characters are contained within a realm the narrator is placed outside of and to which the narrator is, by implication, superior. Privileged access coupled with transcendent knowledge engender a spurious omniscience. The etherealization of the body permits perfect apprehensions: no vagaries of the flesh confound perception. This suppression of the body is accompanied in realistic writing by suppression of the voice. No "marks of enunciation (i.e. the authorial first person)" (Marcus and Cushman 1982, 39) occur in the text. The text purports to become a transparency to the realm of events. Etherealization of the body and transparency of the text tie omniscience to objectivity.

Modalities of Perception

The primacy of vision, to adapt a phrase of James Clifford's (1983, 125), is an aspect of objectivity: vision permits the most remote apprehensions, those in which the perceiver is least involved. Under the aegis of the visual, the body of the Other appears as an object: solid, impenetrable, worn, and worked on the surface, inflected by intermittent movements. The texture of Mrs. Gillette's skin is visible, the flutter of her hands, the tuck of her shoulder, but not what animates these, the sensibility that flushes up under her skin and renders her gestures intelligible, making the tuck of her shoulder, for instance, a flinch (see Geertz 1973, 6–7).[2]

The shift to an internal perspective permits a more proximate perception: hearing. Here, disembodiment leaves behind the ghost of the eavesdropper. As the perceiver moves into the realm of events, the voices of the Others become audible. From this perspective it is possible to catch what is said as well as what is seen. The perceiver begins to haunt the scene just in time to catch the physician's opening line. Fragments of dialogue are suspended between groupings of bodies in space. The presumption of dialogue is that the ethnographer is positioned to overhear the unfolding of meaning between native interpreters. But temporal condensations and expansions occur: skips in the description where something has happened but nothing is textualized; interstices in the description in which something is textualized but nothing has happened. The act of representation becomes an act of interpretation.

The self without the body turns into the narrative voice. That voice can then objectively recount the discourse of the Others. In their discourse, a contrast appears between what Elliot Mishler calls the voice of medicine, used primarily by physicians, and the voice of the lifeworld, used primarily by patients (1984, 63). The contrast is ramifying but it turns on the assumption that the physician takes on an objective discourse whereas the patient retains a subjective involvement in the realm of the ordinary. The difference between voices is clear in this description in the contrast between the patient's concern about the bruise on her left breast and the physician's concern about the tumor in her right one.[3] By speaking objectively, the narrative voice makes itself an analogue of the voice of medicine, implicitly heightening the priority of the physician's perception over the patient's. In aiming for objectivity, realistic writing can thus veer away from it.

The Primacy of the Visual

The suppression of narrative voice heightens the contrast between the self who sees and the Other who speaks. The realm of the audible is transfixed by the realm of the visible. The perspective of the ethnographer contains the voices of the characters. They themselves display no interiority, no subjectivities to which to connect their remarks, only visible objects from which they issue. So the remarks are stuck on, as it were, the wrong way round, tacked onto the outsides of the characters rather than issuing from within. As a consequence, there is no sense in the scene recounted of the production of meaning, rather of a witnessed mystery. The matrix of understandings which connects bodies to remarks is attenuated.[4] Seen from without, the body of the Other is an opaque object to which subjective utterances are attached. An estrangement is created between the insides and outsides of characters, between mind and matter, self and body. The detachment of informing intentions from expressed remarks is a version of the mind/body problem: the problem of how to connect inner thoughts with outer acts. That problem is in part an artifact of the realistic convention. The convention thus dismantles itself: realistic writing makes it impossible to get access to the Other it purports to capture.

The Category of the Other

The congruence between seer and speaker creates a seamless discourse through which the voices of characters erupt. This is the opening of realistic writing toward the Other. But the Other is represented as a character within a realm constituted by the ethnographer as writer. The narrator's voice remains objective whereas the characters speak in their own voices, subject not only to the constraints of their realm but also to the constitutive disposition of its inscriber. Voices split: the subjective voices of natives become estranged from the objective voice of the narrator. This sequestering of voices is itself a claim to objectivity. Inset dialogues in which the characters speak are enclosed in a scholarly discourse which subsumes them. Native voices become exempli of themselves, displays, disclosures, devices of another agency: witness the reification in the text of assumptions of native subjectivity, narrative objectivity. Discursive grounds for native objectivity are undercut by the realistic convention. Herein lies the flaw in the presumption of dialogue: the unfolding of meaning between interpreters in one realm is itself unfolded by another interpreter in another realm. The characters' discourse is captive discourse.

Disembodiment

If the subjective voice of the Other issues from an opaque body, the objective voice of the self inheres in a transparent percipient. The perceiver's transparency takes the form of invisibility, inaudibility, and impalpability in the realm of events.[5] Transparency permits bodiless intrusion into the scene. Characters appear to be unaccompanied in locations where the ethnographer was present (Rimmon-Kenan 1984, 95). The intruder is instantiated in the text as the omniscient narrator. The intent of this effacement of self is to eschew subjectivity; its effect is to disembody the ethnographer.

Alignment and Complicity

The evanescence of the ethnographer's body might be regarded as suspect in realistic writing. A bit of trickery is taking place, a false implication that the ethnographer was not there, not impinging on the scene. But the adumbration of a narrating self out of this absent body is still legitimate. The act of narration can bypass the body and stand apart. The consequent estrangement of the ethnographer from the native permits the ethnographer's alignment with the reader. "The narratee is, by definition," Rimmon-Kenan writes, "situated at the same narrative level as the narrator" (1984, 104). Writer and reader are involved together in the production of a discourse. That discourse becomes the discourse of a discipline. The temptations of this alignment for the reader lie in the reader's entanglement in its putative omniscience. The claims of realistic writing reiterate the claims of scholarly discourse.

The realm of events which contains natives becomes an enclave in scholarly discourse. That realm is not inhabited by natives and discovered by ethnographers, it is mutually constituted between native and ethnographer in fieldwork. In writing, the realm of events is not conveyed by writers to readers but again reconstituted between them. The constitution of a realm is always dialogic. As a writer in the realistic tradition, the ethnographer intrudes not only on a dialogue between natives but also on the dialogue among colleagues. The imprint of the Other in the text is as much the imprint of the narratee as it is the imprint of the native.

The disjunction of the perspective and voice of ethnographer and native in realistic writing permits the conjunction of the perspective and voice of writer and reader. This confluence of perspectives and voices blurs the boundaries between personae. The writing self becomes complicit with the reading other. In effect, *I make myself you.* Complicity aligns us as selves against the native as Other. Our withdrawal from the realm of events fabricates an absence in order to put forward a presence. The absence of the self purports to leave us in the presence of the Other. But the language of objectivity renders the Other an object, the interpreted for which we are interpreters. So what is it I implicate you in? I implicate you in the attempt to bring the Other here, under our eye, as specimen.

Fictionalization

The conventions of perspective and voice compose objectivity in ethnographic writing. To expose these conventions as fictive is to render realistic writing suspect, not because its claims are necessarily epistemologically unwarranted but because it partakes of the literary enterprise. In fabricating a scene, however, this hobnobbing with fictive conventions is inevitable: all scenes entail a perceiver-narrator from whose point of view the realm of events is described. In fabricating realistic discourse, the convention banishes the ethnographer from the scene to recur, disembodied, as the invisible perceiver, the omniscient narrator. Exposing this convention lays the discourse open to the specific suspicion that distortion has entered the ethnographic account in the way the writer constructs that account for the reader: the issue of fictionalization.

Two alignments are available to the ethnographer with respect to this discourse: an alignment with oneself-as-character and an alignment with oneself-as-writer. As character, the realm of events is rendered from an anchorage inside it, returning the self to the horizons of its experience of that realm. As writer, the realm is rendered from an anchorage outside it, in the realm of discourse. Both alignments are legitimate, and legitimate in the same way: by virtue of embodiment. The flaw, if there is one, in alignment with oneself-as-writer lies not in the bodily withdrawal or disembodiment of the ethnographer but in the constraints that positioning puts on access to the Other. One alternative is for the ethnographer to take up the other alignment, with oneself-as-character, in order to produce subjective discourse.

Shifting from objective to subjective discourse allays the suspicion of fictionalization by embodying the ethnographer in the realm of events. In subjective discourse, I align with myself-as-character. The warrant for doing so is the same as the warrant for aligning with myself-as-writer: embodiment. Only this alignment is with another body, not the one I inhere in now, here, but the one I inhered in then, there. Access to that body is still problematic: it is not precisely my body but some permutation, aspect, or incarnation of my body which is lodged in the past.

> *SCENE 2*
>
> *((The examining room is tiny, barely three times the width of its low narrow cot and twice the length. When Dr. English and I come in, Erica Weber is already sitting on the edge of the cot wearing a short white paper jacket that opens in the front and her trousers and shoes. I sit in a chair near the head of the cot. Dr. English sits in the chair beside it so that he and Ms. Weber are almost knee to knee.*
>
> *Unceremoniously, he says, "Let's take a look at it." As she starts to lift the jacket off, he takes hold of the outfolded corner of the shoulder of it with his fingertips, not touching her body. The jacket rustles slightly. Underneath, I see that Ms. Weber is wearing an ace bandage wound around her chest, very pale against her dark skin. She may be pregnant, she hasn't had a period in three months, though she has not yet taken a pregnancy test. I watch as Dr. English unclips the bandage and unwraps it, leaning around her back to gather it up. Her breasts are very dark, heavy and well shaped, set low with large black nipples. Above the right breast is a gauze bandage. Dr. English*

asks Ms. Weber to lie down and takes off the gauze. Along the upper edge of the areola is a deep cut, stitched together loosely so that the skin puckers between the threads. The edge of the wound appears black and not quite closed. I am less aware now of her breasts than of her wound.

Dr. English holds her right shoulder while he stoops to look at the incision. Then he feels around it with the thumbs and fingertips of both hands. I notice that she lies with her arms pressed along her sides and her eyes on the ceiling and wonder how painful the examination is. He says that the only thing they would be worried about now would be if this thing got infected but that that's not likely at this stage. The smell of the room is compounded of disinfectant, sweat, and a sour smell—the wound? Or is it a sour taste at the back of my throat? He holds the skin over the wound with his left fingertip and clips the stitches with scissors in his right hand, bent over her reclining body. I flinch at each clip. He warns her that this may pinch a little bit but she catches her breath with pain. He seems surprised.))

Dr. E: Is it hurting?

Ms. W: Yes.
Still a little tender down there.

Dr. E: O.K.
Well I can see why you feel it's tender.
It does look like you have some infection there.
((Dr. English decides to clean out the wound with swabs dipped in iodine. He sets supplies out on a cloth laid across her lap. He puts on gloves and pinches together the edges of the wound with two fingers of each hand. As her discomfort mounts, I become aware that the edges of my vision are going black and I feel dizzy and queasy. Dr. English balances his left middle finger next to the incision, which gapes a bit, and inserts sterile scissors into it to hold it open. Then he takes a swab in his right hand and probes the wound deeply, bringing out blackish matter and pus. I am aware in my own body of the roughness of the lips of the wound, its tender inner parts, and the depth to which he plunges the swab.))

SUBJECTIVITY

Subjective discourse is embodied discourse; it issues from the body. The access I have to my own body provides the point of insertion for a perceiver in the realm of events. That realm then adumbrates itself around me as its centrality. So centering the self decenters the Other. My body becomes the medium through which intelligence is conducted to the reader.

The anchorage of perception in the body sustains an impression of subjectivity by narrowing awareness to its lodgment in the ethnographer's cranium, just behind the eyes. A deprivileged observer is constituted a participating character. Subjectivity shifts suspicion from my relationship as writer to what is written to my relationship as ethnographer to what has happened. Subjective discourse is open to the suspicion that distortion has been introduced into the ethnographic account through the way the ethnographer interprets the native: the issue of subjectivity. In subjective writing, perspective on what happens is internal from within the body of a character who happens to be myself. This is the perspective known in literary theory as the *internal view* or *narrator-as-character* (Rimmon-Kenan 1984). Events are passed through the body of the perceiver so that the account of them is not reflective but experiential. An intervening sensibility is given priority. The perceiver is positioned in the realm of events at eye level in a bodily lodgment whose orientation and mobility are determined at the caprice of its animator. This perceiver has a single viewpoint with a narrow focal range so that the realm of events is a matter of partial appearances: obtrusions, interpositions, inaccessibilities, and concealments.

Perspective

Space discloses itself to and for the perceiver from within the realm of events. The internal perspective is extended to the perceiver's limited knowledge, emotional engagement, and commensurability with the Other. Temporal perspective likewise unfolds from within the realm of events, spinning out over the course of the scene, not as a rhythm established by the flow of events but as a fluctuation contained or shaped by the perceiver's experience of them. Time is felt time, or durée (Schutz 1973, 253), the sense of time of the perceiver. Epistemological perspective is confined not only to the horizons of the scene but also to the body of the seer. The scene appears stripped of prior and subsequent knowledge. What is known is keyed to and contained by the scene, disclosing itself unevenly over the course of it, so that epistemology is a matter of partial knowledge. The emotional perspective on the scene is engaged rather than detached. The perceiver experiences flushes of feeling in response to the events that transpire. These are indicated by evaluative language about inner states. Emotions are suffered by the self, not attributed to the Other. The ideological perspective on the scene is the character's, not the writer's. Judgments about the nature of reality are judgments by a perceiver implicated in that reality so that her view has an ideological idiosyncrasy.

Voice

From the internal perspective, the scene is recounted by a narrator who has participated in it. Voice is therefore intradiegetic and homodiegetic. This is indicated by such marks of enunciation as first-person pronouns *(I, me, myself)* and laminator verbs *(felt, saw, supposed)* (Goffman 1974, 505). This is the voice called in literary theory the *first-person narrator* or *personal narrator.* Such a narrator is perceptible in the text, indeed insisted upon textually. The perceptibility of the narrator keeps the text from pretending to transparency to another realm by insisting on an intervening sensibility. In subjective discourse, perspective and voice are so connected that the intradiegetic voice issues from the body of a character. The intersection of seer and speaker in the body puts the narrator on a level with the characters. Voice is akin to native voices; perspective is peculiar to the body of the perceiver. On these grounds, the narrator can appear unreliable since, as Rimmon-Kenan points out, "The main sources of unreliability are the narrator's limited knowledge, his personal involvement, and his problematic value-scheme" (1984, 100). Embodiment ties limited knowledge to subjectivity.

Authenticity

Subjective discourse abandons claims of transcendent authority in favor of immanent authenticity. The ethnographer speaks in her own voice from inside her own body. Bodily presence lends ethnographic discourse authenticity, the authenticity of experience. The ethnographer-as-character warrants her discourse by enunciating presence. George Marcus and Dick Cushman describe the gesture of authentication, which they mistakenly elide with authority.

> In fact what gives the ethnographer authority and the text a pervasive sense of concrete reality is the writer's claim to represent a world as only who has known it first-hand can, which thus forges an intimate link between ethnographic writing and fieldwork. Ethnographic description is by no means the straightforward, unproblematic task it is thought to be in the social sciences, but a complex effect, achieved through writing and dependent upon the strategic choice and construction of available detail. (1982, 29; also Clifford 1983, 130)

Authority and authenticity are antithetical. Authority arises either out of the writer's remoteness from the scene she describes or her detachment in the scene she observes; authenticity arises from her investment in description and observation. The cleavage between the two is the result not of lodging a self in or out of the realm of events but of lodging a self in or out of a body. Being in a body, either external to the realm of events (the writer's body) or internal to it (the body of a character) lends the discourse authenticity but impairs its authority;[6] disembodiment enhances authority but abandons authenticity. A discourse is characteristically either authoritative in virtue of appearing impersonal and reliable or authentic in virtue of appearing personal and unreliable. Authoritative texts glean authenticity from brief interpolated subjective passages. The effect of such textual juxtapositions can be in question. Morroe Berger observes with respect to its use in the novel, "The intruding author is now widely held to make the entire relationship between author and reader less credible, yet at the beginning of the novel such interruptions of the narrative were thought to enhance credibility by suggesting that the author could personally attest to the truth of his story" (1977, 174). The thrust of subjective discourse is not to tell *what* I have seen but *that* I have seen. I, the narrator, am embodied in the realm of events.

Modalities of Perception

I materialize in a realm which impresses itself on my own person. My senses become aspects or modes of apprehension of its properties. Tactile apprehensions (as well as olfactory and gustatory ones) require physical proximity. Auditory and visual perceptions can draw progressively away from their subject. Modalities of perception can be ranged on a continuum from proximate to remote: taste, smell, touch, hearing, seeing. In subjective discourse, remote apprehensions give way to proximate perceptions. The visible and the audible are transfixed by the sensible. Vision loses its primacy. Ascendancy has been given instead to senses which are culturally deprivileged as modes of apprehension. Apprehension through touch, taste, or smell is supposed already to be contaminated by interpretation. Even hearing is tinged in this respect. Only the privileged sense, vision, is taken to be inherently free of the bias of the observer, even though here, too, observation is filtered through the body. What D. A. Miller calls "a rigourously enforced separation in the subject between *psyche* and *soma,*" which ensures that "what the body suffers, the mind needn't think" is effaced (1986, 108). Discourse becomes sensational, figuratively as well as literally, and hence suspect.

SENSATIONAL DISCOURSE

The discourse of sensation focuses the subjective account on the affinity between the body of the ethnographer and the body of the patient. Both apprehend events in the realm of medicine experientially. Objective accounts, by contrast, focus on the affinity between the discourse of the writer and the discourse of the physician, both of whom are taken to apprehend events analytically. The ethnographer's bodily affinity with the patient makes the narrative voice implicitly analogous to the voice of the lifeworld and estranges it from the voice of medicine. The difficulties the physician has in the examination are elided in favor of the difficulties of the patient. In an inversion of the objective discourse in which the physician's apprehension of the lump supersedes the patient's experience of the bruise, here the patient's experience of pain obscures the physician's experience of puzzlement. Because ethnographer and native are characters in the same realm, meaning is not overheard, caught by an invisible perceiver, but constituted corporeally among characters of whom the ethnographer is one. In so doing, Miller suggests, meanings disappear into sensations. "Bodies 'naturalize' meanings in which the narrative implicates them . . . Incarnate in the body, [meanings] no longer seem part of a cultural, historical process of signification but instead dissolve into an inarticulate, merely palpable self-evidence" (1980, 108). What is represented in subjective discourse *is* the act of interpretation. This convergence of the perspective and voice of the ethnographer with the perspective and voice of the patient provides a path of access to the Other. The opposition of transparent percipient and opaque object vanishes. Self and Other become jointly transpicuous to the realm of events. Both bodies are suffused with sensation. They disclose signs of presence in the flesh.

The Category of Other

As the ethnographer's body solidifies and the patient's body attenuates, they become permeable. Body boundaries are blurred. Sensations proper to one affect both. The physician cleans the patient's wound; the ethnographer winces. One body takes the impression of what the Other expresses. The interiority of the Other is attested to and expressed by the self. By this device, the mind/body problem is dissolved. The properties of mentality and physicality are distributed over two persons so that sensation is not a private phenomenon but a public discourse. But the distribution of properties is differential: the body of the Other exudes intelligence which is incorporated into the body of the self. Nausea, as Jean-Paul Sartre observed, is the ultimate mark of the Other on the body of the self (1964, 314–315). So the intimacy achieved by this device is with the wrong body. The patient twitches, flushes, sweats, modulates with feeling which is absorbed into the sensibilities of the ethnographer. Feeling is communicated by contagion. What is inscribed on the Other can be read off the self. Access to the Other is again problematic, not because the Other is estranged as an object but because the Other is engulfed as a subject.

As bodies become transparent, the text becomes opaque. Inscription, the imprint of the Other on the body of the self, becomes dialogism, the imprint of the Other on the voice of the self. The ethnographer intrudes not only on the realm of events but also in the dialogue itself. Dialogue takes the form of *free indirect discourse,* the literary convention in which the narrator's voice becomes entangled with the character's (Rimmon-Kenan 1984, 110–116). Narrative voice is itself dialogic. This entanglement of voices extends subjectivity to the Other. But the Other is confounded with the self.

Embodiment

If the intent of embodiment is to sustain subjective discourse, its effect is to reify the body. Recrudescence of the body becomes problematic for subjective discourse just as its evanescence is for objective discourse. In subjective discourse the ethnographer's body accrues visibility, audibility, sensibility. Intrusion into the realm of events is now bodily intrusion. The ethnographer is palpably present to the ethnographic occasion and co-present with body of the Other. Indeed, the body of the ethnographer is so sutured to the Other and the world that it can be neither altogether disentangled from them nor altogether infused into them.

Alignment and Complicity

Subjective discourse invites the reader to align with the ethnographer in the realm of events. The attempt is not to bring the Other here, as specimen, but to take you there, as accomplice. *I make you myself.* The attraction of such an alignment is participation in the modalities of experience. "Our reading bodies [become] theaters of neurasthenia" (Miller 1986, 107). But that is also its flaw. An invitation to experience meets the bodily skepticism of the reader about the ethnographer's experience. The reader is tempted to test her own sensibilities and intelligence against the ethnographer's. In the realistic convention, the narratee accepts the narrator's perspective and voice, but unreliability can cause the reader to sequester the narrative perspective and voice. If objectivity is a discourse of estrangement from the native, subjectivity is a discourse of engagement with the native but an engagement which risks instead estranging the reader.

Estrangement from the reader splits subjective writings off from the objective discourse in which they are inserted. Subjective texts become enclaves of a different order enclosed in scholarly discourse, either in the form of subjective passages within an objective text or of subjective texts within an objective discourse.[7] Objectivity becomes the province of the reader who uses the stance to view, critique, eschew, and estrange the ethnographer's discourse. Narrative voice is heard as a native voice, immediate, unreflective, naive, stuff *for* analysis, not *of* it. Authenticity endangers authority. Subjective writing risks professional repute. I (along with the Other) become the interpreted, you the interpreter.

Subjective Discourse

Subjective discourse, too, disclaims its own fictionalization, not by absenting its narrator but by impersoning her. But my embodiment is as much a fictionalization in subjective writing as my disembodiment was in realistic writing. I reconstitute a realm in which I materialize some past incarnation of myself as a person. In doing so, I make myself a character, as much as, and in much the same way as, I make the Others characters. But I am no longer identical with the self I conjure up. I have fabricated my body along with my discourse (Goffman 1974, 520). Being out of body, as in the objective convention, or being in a past body, as in the subjective one, are equally fictionalizations. A body is merely dematerialized in one and materialized in the other.[8]

One solution for ethnographic writing is not to abandon fictive conventions in order to clear the enterprise of suspicion but openly to engage in fictive modes of inquiry in order to arrive at its center: access to the Other. Neither objective nor subjective discourse ensures access to the Other. Objective discourse forfeits its claim on account of disembodiment of the self: the Other is represented as pure exteriority. Subjective discourse forfeits its claim on account of decentering the Other: the Other is incorporated as pure interiority. Maurice Merleau-Ponty writes:

> The perceiving mind is an incarnated mind. I have tried, first of all, to re-establish the roots of the mind in its body and in its world, going against doctrines which treat perception as a simple result of the action of external things on our body as well as against those which insist on the autonomy of consciousness. These philosophies forget—in favor of a pure exteriority or of a pure interiority—the insertion of the mind in corporeality, the ambiguous relation which we entertain with our body and, correlatively, with perceived things . . . And it is equally clear that one does not account for the facts by superimposing a pure contemplative consciousness on a thinglike body. (1964, 3–4)

I propose a third discourse which moves among the perspectives and voices of Others over the course of the description to create an impression of what might be called intersubjectivity.[9] Shifts in perspective and voice are legitimated by embodiment: the ethnographer's awareness of the insertion of herself in her body provides grounds for access to the experience of other selves in other bodies.

SCENE 3

((Dry, smooth, warm fingertips rest lightly behind my left shoulder as, facing me, Dr. Eden examines with his left hand my right breast and underarm. Though not actually off balance, his touch on my shoulder, counterposing the pressure on my chest, creates a balance turning on the axis of my spine as I sit. The initial shoulder-tip alludes to and foreshadows its counterposed pressure. By virtue of his holding a balance between these two thrusts, of thereby creating in me a sense of being in balance, but one dependent on his balancing me, I entrust my balance to him. Almost, he sets me off balance in order to sustain my balance himself, to reposition the locus of balance in him. Entrusting my body to him then informs the other kind of trust intended in this relationship, my reliance on his expertise.

The kinesic awareness recurrently created and sustained over this phase of the examination by the repositioning of his fingers on my breast or chest accompanied by a slight compensatory shift of the fingers on my back, my awareness of myself as held, suspended, balanced, or inserted in space, is enhanced by the absence of eye contact. He gazes over my right shoulder, head lifted, eyes unfocused, even, sometimes, closed, evidencing that here it is his hands, not his eyes, that receive intelligence. I gaze in a like abstraction over his right shoulder, but manage, on account of his more extensive absorption, to glance occasionally at his face without making eye contact. Thus the circuitry established between us, the kinesics of two bodies in space, their slight inclination together and balancing in respect of each other, these pathways of communication, become dominant, central to the occasion.

This sense of our bodily co-presence is so accomplished that brief releases leave me a little at sea, off-balance, temporarily afloat. Then, Dr. Eden places his left hand behind my right shoulder before continuing the examination of my left breast with his right hand, this positioning of his left hand thus serving to warn me and set my balance for the upcoming leaning of his right hand. If, at the outset, I balanced myself on the edge of the examination table and was put off balance, puzzled a bit, by the first touch of his fingers on my shoulder, now, when left alone I experience my body as a little off-balance and wait to come into balance with respect to his. The touch on the back counterbalances and thus prefigures the subsequent touch on the front.

I am aware initially of the precise texture of Ms. Fielding's skin, fine, smooth, and faintly resistant, mounted flexibly over the curved bands of the ribs, then, at the upper outer corner of her right breast, the thin edge of an expanding plane of intermediary tissue, slightly granular, shifting under my hand over its bony understructure. Under the soft papery texture of the areola and the node of the nipple, the granular tissue thins. I shift my left hand behind her shoulder and feel the distribution of tissue under my right hand against the skeleton. The substance of her body between my hands has a variation of textures which I palpate, delicately with the tips of my fingers, deeply with the pads of my hands. Here the body delivers its intelligence to me kinesthetically. I shift my hold to the other shoulder and begin to mold the tissue from the hollow just under the collar bone to the upper outer edge of the breast, round the outer rim, and across the body of the breast, then to the inner rim where the skin thins and tightens down across the breast bone.

From Dr. Eden's perspective, the pivot of energy is centered on the balls of his feet and his hip joints, from which he inclines forward and back to draw their two bodies into a dynamic balance. He experiences Ms. Fielding's breasts and chest not so much as a surface whose contours are to be described by his fingertips but rather as a solid, a shape whose pliabilities and resistances are felt as a thickness or density of the entity between his hands, felt as substance and volume rather than surface, something more akin to sculpture than cartography.

Dr. Eden does not use much cover patter, the magician's trick for directing attention strategically away from what his hands are doing. Nor does he allude to what is at hand. But his touch of her shoulder comes to operate metacommunicatively to create the expectation of a next move, one of a narrow class of possible next moves. The examination of the chest is sequential to the examination of the back so that his touch on Mrs. Fielding's shoulder, though puzzling in its import when initiated from the front, is not surprising. Her body is already accustomed to being touched there. Indeed, the examination of the back, neutralized even more by the intervention of a stethoscope, precedes and modulates to the examination of the front, thus creating a context of neutrality for that touchier matter. Doctor and patient are not so much touching each other as in touch.))

INTERSUBJECTIVITY

In intersubjective writing, a realm of events is disclosed from the perspectives of its inhabitants, whose voices are audible and whose bodies are palpable. The Other is rendered a self. In consequence, relationships between perceiver and perceived, narrator and narratee, ethnographer and native, writer and reader, are transformed. The perceiver sees from other bodies, the narrator speaks in other voices, but they are ours, we are in them. The question at issue becomes what can scrupulously be said from the perspective of the Other, in the voice of the Other.

Perspective

Perspective in intersubjective writing is internal to the realm of events and from either within or without the bodies of characters. From within, space adumbrates around the body of the character as its centrality, time takes its rhythms from the body, and knowledge discloses itself to and in respect of, the body. Emotion, though focalized from within the body, retains an external perspective, as if the characters reflected on rather than underwent the experience of their own bodies. Ideology, likewise focalized from within the body, also suggests an external perspective. It puts forward the sort of thing the body is rather than the character's idiosyncratic notion of it. This sense of the body is slightly differently focalized by each character. The patient feels as if her body were all surface. She is aware of variations in texture, pressure, sensitivity. The physician feels through the surface to the understructures, as if, on the contrary, the body were all form. But both senses are rooted in a conception of the body as a malleable substance transmitting intelligence to and from the body of the Other.

Spatio-temporal perspective shifts in the scene from a locus within the body of one or the other character to a locus without the bodies of characters but inside the realm of events. From there a third perceiver, disembodied, hovers without the bodies of characters but still receives intelligence through them. Space wraps itself around the characters, taking then along bodily, and time unfolds to their body rhythms. But knowledge, emotion, and ideology, though focalized from within the body, retain an extension, detachment, and objectivity which transcend the body. Indeed, it is the epistemological, emotional, and ideological perspective of the disembodied observer which the characters seem to have appropriated.

The shift of spatial perspectives among embodied and disembodied perceivers presents multiple footholds in the realm of events. For each perceiver, time moves sequentially through the events of the scene but from the perspective of the scene as a whole, time moves back over the same sequence from different perspectives. What is climactic in the other scenes is recursive here. In the first scene, the climax is located from without the body, the disclosure of the tumor; in the second, it is located from within the body, the access of nausea. The structure of climax is part of what holds the reader's interest in the scene: suspense spins out retrospectively from the climax to its preliminaries. The third scene unfolds and refolds spacetimes to produce a laminated description of the body and its surround from within and without. Despite its lack of theatricality, the third description, too, seduces us, I would argue, by its involvement of the body. Its structure can also be subsumed under a sexual metaphor, but a feminist one of diffuse stimulation rather than a masculinist one of focal climax. Knowledge is embodied knowledge.

The impression of intersubjectivity in this passage is heightened by loci in the bodies of Others and by shifts of perspective among percipients but it can be achieved from one locus in one body by attention to inscription: the imprint of the Other on the body. Such writing constitutes the body the medium through which intelligence of the Other is conducted.

Voice

The narrating voice is rendered a voice of the self issuing from within or without the body. From within the body, the voice belongs to a character who participates in the realm of events and so is intradiegetic and homodiegetic. The narrator is perceptible in the text in marks of enunciation: personal pronouns (I) and laminator verbs (aware, look, feel, expect). Yet the narrator does not seem to be unreliable. The body interposes, here, a sensory apparatus rather than a sensational apprehension. In this capacity, it takes the imprint of the other.

Behind, beside, beneath the voices of characters another voice is hidden, the disembodied voice of a narrator who articulates, imperceptibly and reliably, events in which the narrator does not participate. This voice is technically extradiegetic, speaking from beyond the realm of events, but it becomes entangled with the intradiegetic voices of characters so that it is difficult to decipher who speaks for or as whom. The result is technically free indirect discourse of the sort known as indirect interior monologue (Rimmon-Kenan 1984, 114). Not only are the voices of the characters audible in the voice of the narrator from without but also that narrator's voice inflects theirs. Who is it that reflects on balance, on texture, on kinesthetics? The physician? The patient? The ethnographer? If the disembodied voice is dialogized, the embodied voices are authorized. The shift of voices between embodied and disembodied narrators produces the pattern of what Marcus and Cushman (1982, 43), following James Clifford (1983, 140–142), call "dispersed authority."

The impression of intersubjectivity is enhanced in this passage by multiple voices and shifting among narrators but it can be achieved in one voice by attention to dialogism, the imprint of the Other on the voice of the self. Such writing constitutes the voice the medium through which intelligence of the Other is articulated. Inscription and dialogism are the marks of intersubjective writing. Inscription, the imprint of the Other on the body, matches dialogism, the imprint of the Other on the voice. Writing the body of the Other, then, invites multilocality just as writing the voice of the Other invites multivocality.

Authority and Authenticity

The complex of holds in intersubjective writing yields an admixture of authority and authenticity. The voices of embodied characters, which might be heard as authentic, are positioned to pick up some of the accents of authority from their flashes of external perspective. The voice of the disembodied perceiver, in contrast, would be authoritative but for the audibility through it of other voices, the perceptibility in it of other bodies.

Intersubjectivity recovers from subjectivity, embodiment, and from objectivity the centrality of the Other. Perceivers are present in a realm of events which adumbrates itself around them as its centrality. Experience is inscribed on the body. But the body is not the body of the self but the body of the Other or, rather, the Other rendered a self. Intersubjectivity loosens perception from its lodgment in the ethnographer's body to insert it into the body of the Other. Authenticity, deriving from embodiment, is relocated in the body of the Other. Authority, deriving from disembodiment, is dismantled and reconstructed in the voice of the Other.

Modalities of Perception

The hierarchy of perceptual modalities remains inverted in intersubjective writing. Vision, the culturally privileged modality, is subordinated to sensation.[10] Audition is banished from the scene. Or, more precisely, the locus of the audible shifts from without to within the body. Dialogism, the public locus of discourse, becomes interior monologue, its embodied locus. Embodiment is the pivot of intersubjective writing. Hence such writing moves remote apprehensions toward proximate perceptions, so that they display an affinity with the culturally deprivileged modality, touch.

Touch is suspect because information which has passed through the body of the perceiver is regarded as idiosyncratic, peculiar to that individual. The reembodiment of seeing and hearing lays them open to the same suspicion. However, in this discourse, the felt body is presented as a sensory perceptor on which information is precisely and delicately inscribed. But the import of the information for its percipient is not presupposed. Indeed, that is the very question on which intersubjective discourse focuses attention. Without assuming either universality or idiosyncrasy of emotions, a nexus of circumstances is presented, in or toward which emotions are constructed or directed. In the presentation, sensation is disentangled from emotion. The body takes impressions rather than imposes interpretations. This abstinence separates intersubjective writing from subjective writing.

The Category of the Other

Bodily access calls into question the category of the Other in ethnographic writing. Objective writing sustains a view of the Other as estranged. The realm of the other is represented as remote from the ordinary and the other is inscribed as exotic. Subjective writing, on the contrary, attains an intimacy with the Other. The realm of the Other is represented as an adumbration of the ordinary and the Other is inscribed as familiar.[11] Intersubjective writing puts forward the body as the hold on access to the Other. We are moved to consider the incorrigible estrangement of the familiar Other, the uncomfortable intimacy of the exotic Other.

Embodiment and Disembodiment

In intersubjective writing, as in objective writing, the body of the ethnographer is banished from the scene. But as the body dematerializes, the self is relocated in the body of the Other. In intersubjective writing, as in subjective writing, I am embodied, but not in my own body. I intrude in the body of the Other.[12] Intrusion is neither bodiless nor bodily but embodied. The evanescence of my body is commensurate with the recrudescence of the Other. Thus the Other becomes an embodied self.

Alignment and Complicity

Abstinence in intersubjective writing extends to abstinence from claims about inner states. All modes of touch described, though presented from within the bodies of characters, can be observed from without the bodies of characters. The focal point of perception has simply been pivoted into the body of the Other without altering the content of the observations, with one exception: a sentient awareness of touch has been attributed to the characters. They are presumed to be alert to what transpires. The patient's apprehension of texture or pressure, the doctor's sculptural modeling of the body, become part of their discourse. Here the ethnographer's knowledge, knowledge of these characters in this scene, of the constituent properties of such scenes in medicine, and of the ontological conditions of medicine are reconstituted in the Other. Aspects of the account, which in objective writing are separated out from the description as the analysis, are returned to the status of situated knowledge. Specifically, the difference between one's ordinary sense of one's body and the specific twist medicine imparts to it are represented discursively in the sensibilities of these inhabitants of the realm of medicine.

The ethnographer in this style of writing is aligned with neither adumbration of herself, as-character or as-writer, but with the Other. The locus of the perceiver in the realm of events and of the narrator in the realm of discourse offers a foothold or discursive position to the reader. The reader is invited to align, like the ethnographer, with the Other-as-character. This invitation is neither undercut, as in subjective writing, by the imposition of the sensibilities of the ethnographer nor, on the other hand, is it insulated, as in objective writing, from the realm of events. The reader can thus skirt both sympathy, the effect of being swept into and by the experience of the Other, and detachment, the effect of being estranged from that experience, to embrace empathy, the effect of being located in the experience. The ethnographer and the reader are each shifted one lamination deeper into the realm of events: the self becomes the Other, the reader becomes a self. By this gesture, *I make us both the Other.* The reader is perceptually positioned to perceive the experience of the Other and discursively positioned to hear the voice of the Other.

And the thrust of that move in ethnographic writing is to *tell* what the Other has *seen.*

FICTIONALIZATION AND SUBJECTIVE DISCOURSE

Claims of access to the experience of the Other have been suspect in ethnographic writing. What are the grounds of this suspicion? On the objective view, access is impossible. The mind is taken to be so secreted in the body that embodiment itself foils access. The consequence of that is the discorporation of the Other as an exotic object whose external properties might be *seen.* On the subjective view, access is held in some sense to be possible but idiosyncratic. The mind is supposed to be so immaterial that it transcends the body. The consequence of that is the incorporation of the Other as an intimate subject whose internal properties might be *felt.* Both discourses spin out the Cartesian dialectic that underlies them. The estrangement of mind from body is literalized in objective discourse in the disembodiment of the ethnographer. A rarification of the self addresses a concretion of the Other. In subjective discourse, by contrast, the estranged mind of the Other comes to be lodged in the body of the self so that two rarified selves intermingle.

Positioning subjectivity and objectivity with respect to each other polarizes what is not only not antithetical but probably not even separable. Reality is conceived objectively as pure substance, subjectively as the absence of substance. The incommensurability between the two is the root of the mind/body problem: how is subject hooked to object, mind to body, self to substance? The two discourses do not properly occur at the same level of analysis. They work as alternative descriptions of the same event at different levels of analysis. Subjectivity construes events in terms of sense impressions; objectivity gives priority to independently existing objects. Either has difficulty from its starting point in arriving at the other order of events. Given sensations, impressions, ideas, inner states, how do we get to external objects? Given external objects, how do we get to inner states? Events can be construed either as perceptual phenomena whose objective status remains problematic or as material objects whose subjectivity is incalculable. Set together, the two discourses heighten each other's peculiarities. Objectivity excludes consciousness from the body. The body is regarded as an object whose peculiar surface does not necessarily evidence an inner reality. Subjectivity excludes the materiality of the mind. Dislocated from the body, consciousness becomes unbounded, metaphysical, ghostly.

The nexus of subjectivity and objectivity in the body evidences their inextricability.[13] Here the self is materialized; the body is inspirited. The body is not the boundary between persons but the connection, the site, the instrument of their co-presence. The trick is to eschew the objecthood of embodiment in order to apprehend instead both the materiality of the subject and the consciousness of the object. The self is neither magically secreted inside the body nor does it hover spectrally around it. Rather it is suffused through it and expressed out of it. Nothing supernatural is required to get access to the Other.[14] Our relationship with the Other is already replete with presence. The Cartesian dialectic to which we are heir accustoms us to the notion that the body of the Other *conceals* a self. In reality, the body of the Other *reveals* a self.

The root problem in ethnographic writing is not getting access to the Other but so constituting the category of the Other that getting access is problematic. Discourse is incorrigibly tinged with the presence of the Other. Indeed on the Bakhtinian view, the problem would be to avoid access. Objective writing might be said to be designed to do just that by fictionalizing an absence. But if the conventions of objective writing are fictive, then we cannot attain epistemological impeccability by preserving the perspective of the omniscient narrator. In subjective writing, the ontological grounding of description in the ethnographer's insertion in the scene rescues impeccability but blurs the discretion between self and Other. The move out of the objective into the subjective turns on embodiment. If we grant that the body is the hinge of experience, then this move discloses the ontological grounds for intersubjectivity. The devices of narrativity uncover the locus of the Other in the self. Its project is to make us present to the scene bodily but not in our own body. The focal point of perception is simply pivoted into the body of the Other without altering its content. Hence the fictionalized presence in intersubjective writing: the Other-as-self.

In the traditional Japanese folktale "Rashomon" (Akutagawa 1970, 19–34), three characters, including the ghost of one who died, describe the same scene, fabricating a description to suit their own purposes. Each description conflicts with the others but accounts for some of the aspects of the event described so that it is possible, at least for a traditional Japanese judge, to cobble together the truth of the matter. It might be argued that each of these three discourses, the objective, the subjective, and the intersubjective, produce different, though not necessarily incommensurable, perspectives on the Other. Interlayering perspectives does not converge on the truth; the perspectives diverge toward alternate realities. The question is: How are we bound to these perspectives? How do we move among them? How do we legitimate these moves?

• • •

NOTE ON TRANSCRIPTION

NOTES

REFERENCES

INDEX

NOTE ON TRANSCRIPTION

• • •

The difficulty with transcriptions is always what Korzybski calls the map-territory problem (in Bateson 1972, 180). As the map approaches the detail and complexity of the territory it depicts, it becomes increasingly cumbersome to use. Maps are intended to abstract, schematize, and simplify aspects of a topography they do not fully capture and never exhaust for the purpose of a particular sort of traversal. There is no ideal transcription; only different transcriptions with more or less detail of one sort or another. Since no transcription can perfectly represent its reality, each transcription might as well be suited to its purposes.

The transcription for this study is designed to preserve certain aspects of audible communication. In so doing, it inevitably erases others. For instance—a device of Dennis Tedlock's (1978)—line ends indicate pauses, thus eliminating the ambiguous comma, which can mark either a conceptual unit or a spoken one. Equal signs mark the absence of obligatory end pauses, and slant lines mark one-beat pauses, the length of time during which conversationalists will feel that speakers could have spoken up. I never note, as some transcribers do, the exact length of time elapsed, as I consider the significance of pauses a matter of felt time rather than clock time.

Punctuation, too, is oriented to hearing rather than writing. Capital letters indicate the start of utterances, periods indicate down intonations and question marks up intonations at the ends of utterances. Dashes indicate that a speaker interrupted an utterance, usually his or her own, in order to correct it. Parentheses indicate that the transcriber is not sure what was said; anything written inside one is purely speculative, except for laughter, noted as (hehe). Some researchers use capital letters to indicate loudness, miniature letters to indicate whispers, and repeated letters to indicate prolongation. Here, such metacommunications as I deem relevant, including volume or laughter, gestures, bodily orientation, or spatial arrangement, are noted in double parentheses as editorial comments. Double left-hand brackets indicate two speakers speaking at once and single brackets within the text mark the extent of simultaneous speech.

Apart from these peculiarities, the transcriptions work rather like a dramatic script, speakers' initials to the left, utterances to the right, and stage directions in double parentheses. Elisions in the texts are noted with ellipses. English spelling indicates English speaking. I made no attempt to suggest vernacular speech or regional pronunciation, because to do so privileges the unmarked category which, after all, has an accent, too. Phonetic transcription, which does not privilege one speech register, labors the business of pronunciation beyond my purposes.

A summary of these devices follows. Most of them were adapted for me by the linguist Malcah Yeager from James Shenkein (1978).

TRANSCRIPTION DEVICES

Line ends	Pauses
=	Absence of obligatory end-pause
/	One-beat pause
Capital letters	Start of utterance
.	Down intonation at end of utterance
?	Up intonation at end of utterance
—	Correction phenomena
( )	Doubtful hearings
(hehe)	Laughter
(( ))	Editorial comments
[[	Simultaneous speech
[ ]	Extent of simultaneity
. . . .	Elisions

NOTES

• • •

INTRODUCTION

1. Proxemics is Edward T. Hall's term for the study of personal space (1969, 1).

1. DISEMBODIMENT: INTERNAL MEDICINE

1. The sense of protection afforded by such architectural features diminishes as their distance from the body increases: it is exceedingly difficult to feel concealed in a gymnasium and easy in a closet.

2. This view is at odds with that of the post-Darwinians, who take bodily forms and expressions to be ritualizations of physiological conditions and responses and so inherent in the body (Darwin 1969, 28–29). Subtle accounts of this controversy are provided by Erving Goffman (1979, 69); by Richard Wollheim in terms of expression and correspondence theories, and iconicity and arbitrariness (1968, 26–29, 104–107); by Gregory Bateson in terms of primary and secondary process, mood-signs and signals, and digital and analogic communication (1972, 135, 178, 372–374; and by Rodney Needhan in terms of inner states and external expressions (1981, 53–71).

3. This follows Victor Turner's argument for rituals as transformative (1980, 161).

4. My husband, a physician who used his own name, as I do mine, was, under the usual assumptions, occasionally addressed as "Mr. Young." He used to correct the title rather than the name since, as he once remarked, "It's not the 'Young' I object to, it's the 'Mr.' "

5. The notion of being sick as a role derives from Talcott Parsons. "To be sick [is] not only to be in a biological state . . . but requires exemptions from obligations, conditional legitimation, and motivation to accept therapeutic help. It [can] thus, in part, as least, be classed as a type of deviant behavior . . . socially categorized as a

kind of role" (1964, 332). The theatrical metaphor for forms of social life has been neatly explored by Goffman (1959, 240–254) and Natanson (1970, 6, 167).

6. Apparently both the fabric and the form of these coats changes with the status of the practitioners so that discriminations among doctors, nurses, receptionists, technicians, cooks, and cleaners, as well as among medical students, interns, residents, and staff physicians, and between any of these and patients, are indicated by slight modulations of costume. Roughly, the more extensive the garment, the more complete the investment of its wearer in the realm of medicine. Internists wear long-sleeved, three-quarter-length white cotton coats with their last names and title embroidered on the edge of the upper left-hand pocket; their receptionists wear short-sleeved white nylon tunics; their patients wear white paper jackets or gowns. Thus along with realm-shift, a hierarchic organization of statuses is effected.

7. Nurses participate in the access we grant physicians to our bodies. Of course, we grant limited access to our bodies to a range of practitioners from shoe salespersons, manicurists, and hairdressers to dentists, masseuses, and most intimately, nurslings. But apart from physicians and lovers, access to the anal-genital region is specific to morticians and prostitutes, which suggests something about the body taboos that attend such access.

8. Elliot Mishler points out that the conventions could be shifted by the physician as well as by the patient (1984, 128).

2. PERCEPTUAL MODALITIES: GYNECOLOGY

1. Dr. Anna Copperfield described a photograph of an ultrasound of a patient's viscera to the patient, thus making the woman's interior terrain yet another degree more alien from her experience of her body (June 4, 1987, transcript, 12–13).

2. Suggestively, Hubert Dreyfus, following Heidigger, characterizes nursing as a caring rather than curing practice in his introduction to *Interpretive Phenomenology: Embodiment, Caring, and Ethics in Health and Illness* (1994, ix).

3. The gynecologist I did the most work with happened to be male so these gestures are also gendered. From brief observations, a similar relinquishment of control by the patient attends examinations by women gynecologists but it is less apt to be materialized in these body holds and is altogether less marked.

4. Feminists sometimes dismantle the hierarchy implicit in these arrangements by eschewing the assumptions of gowns and drapes that mark difference and insisting on commensurate participation in the interchange.

5. Dr. Copperfield uses a similar formula, "Going to touch you now?" as her characteristic opening move toward a vaginal examination.

3. DECIPHERING SIGNS: SURGERY

1. I can grant priority to neither the subject nor the object. Althusser writes that the great debate in philosophy has always been between idealism and materialism. I agree. And, like Althusser, I am not an idealist; unlike Althusser, neither am I a materialist. For me, the relationship between idealism and materialism remains at issue, the issue, as Althusser points out (1971, 50), always being which takes primacy.

2. Nurses customarily wear knee-length tunics of the same cotton material. When women began to practice surgery, they, too, wore the knee-length tunic. Nowadays, most women surgeons wear tunic and trousers and some nurses do, too. Men nurses never entertained such ambiguities. Like their gender group, men, and unlike their professional group, nurses, they always wore trousers.

3. Originally, women's caps were made full, like shower caps, men's like skullcaps. The distinction between them was gender specified by putting flowers on the women's. The era during which men wore long hair and therefore required fuller caps made this custom less apt. Women, especially women who are surgeons, can now wear skullcaps provided their hair is either quite short or quite long and tied back, as can men. A further cover is provided for the still gender-specific beard.

4. Ricoeur writes, "Still another way of expressing the same enigma is that as an individual the text may be reached from different sides. Like a cube, or a volume in space, the text presents a 'relief.' Its different topics are not at the same altitude. Therefore the reconstruction of the whole has a perspectivist aspect similar to that of perception" (1971, 548–549).

4. STILL LIFE WITH CORPSE: PATHOLOGY

1. This corpse, disgorging its cornucopia of organs, is arguably an instance of what Mary Russo calls the female grotesque (1995). Not only is the medical body, in respect of its passivity and abjection, feminized, but also the female body is the root instance of the grotesque body, epitomized by Bakhtin's "senile, pregnant hags." He writes, "This is a typical and very strongly expressed grotesque. It is ambivalent. It is pregnant death, a death that gives birth. There is nothing completed, nothing calm and stable in the bodies of these old hags. They combine a senile, decaying and deformed flesh with the flesh of new life, conceived but as yet unformed. Life is shown in its two-fold contradictory process; it is the epitome of incompleteness. And such is precisely the grotesque concept of the body" (1984, 25–26). "Moreover," Bakhtin notes, "the old hags are laughing" (1984, 25). "The word itself," writes Russo, ". . . evokes the cave—the grotto-esque. Low, hidden, earthly, dark, material, immanent, visceral. As bodily metaphor, the grotesque cave

tends to look like (and in the most gross metaphorical sense be identified with) the cavernous anatomical female body" (1995, 1). I owe this insight about the grotesque as female to one of the anonymous readers of this manuscript.

2. The medieval body can be a site of representation for the ethereal as well as the grotesque discourse, in the persons of saints and ascetics, for instance, though these embodiments do not necessarily rarify the flesh. Carolyn Walker Bynum writes that both men and women somatize religious experience in the Middle Ages (1989, 167–168) but that women's bodies tended toward asceticisms and ecstasies, men's toward profound stillness experienced as bodilessness (1989, 171, 169). Even these embodiments, as a medievalist historian has pointed out to me, are but a few of the multiple medieval bodies, "probably considerably more than in the modern period when western medicine and mass culture have acted to homogenize perceptions and self-perceptions of the body" (1995, pers. comm.). Nevertheless, it is the grotesque body, even if we have fabricated the medieval body as a historical repository of the grotesque, that is suppressed by medical discourse.

3. The pig taboo in Jewish tradition was enhanced by the animal's reputation for eating refuse but it is rooted, according to Mary Douglas, in the pig's categorical transgressions: "cloven hoofed, cud-chewing ungulates are the model of the proper kind of food for a pastoralist" (1970, 69). "The dietary rules," she writes, "should be taken as a whole and related to the totality of symbolic structures organizing the universe. In this way the abominations are seen as anomalies within a particular logical scheme" (1973, 60). Within this order, an analogy is proposed betwen the altar and the table, the meal taking the form of a slightly attenuated but still pure version of ritual sacrifice. Just as the blemished or impure cannot enter the temple, so the anomalous cannot be eaten. Clearly, "the purity in question is the purity of the categories. . . . The sanctity of cognitive boundaries is made known by valuing the integrity of physical forms. The perfect physical specimens point to the perfectly bounded temple, altar and sanctuary" (1984, 269).

4. Mary Douglas, too, notes the affinity between pigs and death. The boundaries between categories are preserved to maintain the intellectual coherence of a system of thought and, by extension, a body of people. Both pigs and death transgress the boundaries between the sacred and the profane, the proper and the taboo, the pure and the flawed. "Much of Leviticus is taken up with stating the physical perfection that is required of things presented in the temple and of persons approaching it. The animals offered in sacrifice must be without blemish, women must be purified after childbirth, lepers should be separated and ritually cleansed before being allowed to approach it once they are cured. All bodily discharges are defiling and disqualify from approach to the temple. Priests may only come into contact with death when their own close kin die. But the high priest must never have contact with death" (1970, 64).

5. The term "discourse," as it has come into use in the social sciences, does not mean texts, writings, or utterances, *particularly*. "Discourse," I take it, alludes to "universe of discourse," the realm of events, the world, of which texts, writings, utterances, and also gestures, acts, thoughts, spaces, times, objects, and so on are a part. Physicians' way of writing within their own discourse embodies its ontological characteristics. Such writings are not to be regarded as descriptions of a reality but as something more like manifestations, instances, of it.

6. For the technical aspects of perspective and voice, I am indebted to Shlomith Rimmon-Kenan's work (1984).

7. This line of thought from Kristeva was suggested to me by Janet Langlois.

CONCLUSION: THE MAKING OF THE MEDICAL BODY

1. This reflection on corporeal metaphors has benefited from the insights of students in my 1991 course at New York University, "The Social Construction of Body and Self."

CODA: PERSPECTIVES ON EMBODIMENT

1. As Dan Rose writes, "The objects that are formed by ethnographic study have historically been referred to as *natives*" (1989, 320).

2. To describe what the patient did in terms of the contracture of the muscles, the displacement of the bone, the withdrawal of the body, would be to contribute to this opacity. The word "flinch" locates intentionality in the gesture.

3. The convention in medical anthropology, sociology, or folklore has been to look, as Mishler does, at disparities between understandings physicians have of what transpires in medicine and the understandings patients have. This is considered a matter of cognitive dissonance. But the investigation of disparity has gone on without looking at congruity. The intention here is to consider what patients and physicians do together, their mutual construction of the realm of medicine in which they jointly participate, as the ground against which to define their differential contributions. See West 1984, 155.

4. It is, in part, this absence of understandings, in dialogue or in any other form of life abstracted from its contexts, that creates the space into which an ethnographic account necessarily and properly falls.

5. The ethnographer's implied invisibility, inaudibility, and impalpability are sustained by fieldwork practices in which ethnographers behave as if there were not there. Not only do they undertake to represent scenes as they would be in their own absence but even to experience them in that way.

6. It would be possible to argue a hidden inauthenticity in subjective discourse in eschewing the writerly body. But of course it is impossible to be true to both embodiments at once.

7. You may note that this writing itself proceeds by inserting textual enclaves of a dubious character into an ongoing objective discourse. This device puts forward these enclaves as inhabitable realms, presentations of access to the Other, while actually framing them off within the sort of discourse scholars find it respectable to write. So this effect, too, is spurious, engagement set off by detachment. What would the effect be if an alternate discourse were presented without frame as an ethnographic writing?

8. Even being in my own body as a writer would be a fictionalization. Here I would constitute my writing self a character in its own words.

9. Other attempts to characterize a triad of discourses of which the first two are something like objective and subjective diverge in the third, usually a discourse of some degree of idiosyncrasy invented by the author. See, for instance, John van Manaan on what he calls realist, confessional, and impressionist styles of ethnographic writing (1988, 7, 45–124).

10. To frame the scene visually would be to direct attention to a different set of properties, for instance, that the interaction transpires between a middle-aged white woman and a young black man.

11. Marie-Laure Ryan suggests a way to investigate the contrast between extending and traversing boundaries in terms of whether or not what she calls the illocutionary boundary (that is, the boundary between forms of discourse, for instance, between stories and the conversations in which they occur or here between scenes and the ethnographies in which they are enclosed or between an ethnography and the scholarly discourse in which it is an enclave) is also an ontological boundary (that is, a boundary between realities). Is the realm conjured up by ethnographies a separate reality or is it continuous with this one? Are ethnographers in their personae as fieldworkers and their personae as writers in the same realm? The convention for true stories, she argues, is that they are. They cross neither boundary but present the same speaker in the same reality system (1991, 176). I argue that they are not. The realm of the ordinary itself consists of multiple realities. Entering any of them entails crossing an ontological boundary, experienced, for instance, as a modulation of embodiment. The realm conjured up by the text can never be the realm in which the text inheres. As invoked by narrative, the realm of the ordinary is always an alternate reality (Young 1986, 186–187, 197–199).

12. Gelya Frank makes a notable attempt on the body of the Other in her fabrication of the story of a paraplegic (1981).

13. Richard Zaner writes, "Nor is it accidental that both idealism and materialism, each in its own way, leave something central out of joint, something incom-

prehensible which mutely but insistently bespeaks its unaccountable presence: *the living body*" (1981, 7). " 'Body' seems a wonderful enigma indeed (as Pascal acidly noted), within the very system of the dualism: shifting, equivocal, at once *both* 'mental' *and* 'material', yet *neither* the one *nor* the other simpliciter" (1981, 10).

14. Phenomenologists hold that our sense of the presence of another self in another body is inherent in our sense of ourselves in our own bodies: "whenever I try to understand myself the whole fabric of the perceptible world comes too, and with it the others who are caught in it" (Merleau-Ponty 1964, 15).

REFERENCES

• • •

Akutagawa, Ryunosuke. 1970. *Rashomon and Other Stories.* New York: Liveright.

Althusser, Louis. 1971. *Lenin and Philosophy and Other Essays.* Trans. Ben Brewster. New York: Monthly Review Press.

Armstrong, Robert Plant. 1971. *The Affecting Presence.* Urbana: University of Illinois Press.

——— 1975. *Wellspring: On the Myth and Source of Culture.* Berkeley: University of California Press.

Bakhtin, Mikhail. 1984. *Rabelais and His World.* Trans. Helene Iswolsky. Bloomington: Indiana University Press.

——— 1985. *The Dialogic Imagination.* Ed. Michael Holquist. Trans. Caryl Emerson and Michael Holquist. Austin: University of Texas Press.

Barker, Francis. 1984. *The Tremulous Private Body.* London and New York: Methuen.

Bataille, Georges. 1985. *Visions of Excess: Selected Writings, 1927–1939.* Ed. and trans. Allan Stoekl with Carl R. Lovitt and Donald M. Leslie, Jr. Minneapolis: University of Minnesota Press.

Bateson, Gregory. 1972. *Steps to an Ecology of Mind.* New York: Ballantine.

Berger, John, and Jean Mohr. 1976. *A Fortunate Man.* New York: Pantheon Books.

Berger, Morroe. 1977. *Real and Imagined Worlds: The Novel and Social Science.* Cambridge, Mass.: Harvard University Press.

Bosk, Charles. 1983. *Forgive and Remember: Managing Medical Failure.* Chicago: University of Chicago Press.

Bourdieu, Pierre. 1989. *Outline of a Theory of Practice.* Cambridge: Cambridge University Press.

Bruner, Jerome. 1986. *Actual Minds, Possible Worlds.* Cambridge, Mass.: Harvard University Press.

——— 1990. *Acts of Meaning.* Cambridge, Mass.: Harvard University Press.

Bynum, Caroline Walker. 1989. "The Female Body and Religious Practice in the

Later Middle Ages." In *Zone: Fragments for a History of the Human Body,* ed. Michel Feher, with Ramona Naddaff and Nadia Tazi. New York: Urzone.

Carroll, Lewis. 1989. *Through the Looking Glass.* New York: Grosset and Dunlap.

Clifford, James. 1983. "On Ethnographic Authority." *Representations* 1:2, 118–146.

Conrad III, Barnaby. 1986. "Let Them Eat Bread." *Connoisseur.*

Corbin, Alain. 1986. *The Foul and the Fragrant: Odor and the French Social Imagination.* Cambridge, Mass.: Harvard University Press.

Delafield, M.D., Francis. 1972. *A Handbook of Post-Mortem Examinations and of Morbid Anatomy.* New York: Wood.

Department of Pathology. 1989–1990. *Autopsy Manual.* "University Hospital."

Darwin, Charles. [1879]1969. *The Expression of the Emotions in Man and Animals.* New York: Greenwood Press.

Derrida, Jacques. 1973. *Speech and Phenomena.* Trans. David. B. Allison. Evanston, Ill.: Northwestern University Press.

Descartes, René. [1641] 1962. *Meditations.* In Walter Kaufmann, ed., *Philosophic Classics: From Bacon to Kant,* 28–96. Englewood Cliffs, N.J.: Prentice-Hall.

Douglas, Mary. 1970. *Purity and Danger: An Analysis of Concepts of Pollution and Taboo.* Middlesex: Penguin.

——— 1973. *Natural Symbols: Explorations in Cosmology.* New York: Random House.

——— 1984. "Deciphering a Meal." In *Implicit Meanings.* London: Routledge and Kegan Paul.

Duden, Barbara. 1991. *The Woman beneath the Skin: A Doctor's Patients in Eighteenth-Century Germany.* Trans. Thomas Dunlap. Cambridge, Mass.: Harvard University Press.

Feher, Michel, with Ramona Naddaff and Nadia Tazi, eds. 1989. *Zone: Fragments for a History of the Human Body* 1, 2, and 3. New York: Urzone.

Foucault, Michel. 1975. *The Birth of the Clinic.* Trans. A. M. Sheridan Smith. New York: Vintage Books.

Frank, Gelya. 1981. "Mercy's Children." *Anthropology and Humanism Quarterly* 6:4, 8–12.

Geertz, Clifford. 1973. *The Interpretation of Cultures.* New York: Basic Books.

Genette, Gerard. 1983. *Narrative Discourse: An Essay in Method.* Ithaca, N.Y.: Cornell University Press.

Gergen, Kenneth. 1986. *If Persons Are Texts.* New Brunswick, N.J.: Rutgers University Press.

Gergen, Kenneth, and Mary Gergen. 1983. "Narratives of the Self." In *Studies in Social Identity,* ed. T. R. Sarbin and K. E. Scheibe. New York: Praeger.

——— 1986. "Narrative Form and the Construction of Psychological Theory." Swarthmore, Penn.

Ginzburg, Carlo. 1980. "Morelli, Freud, and Sherlock Holmes: Clues and the Scientific Method." *History Workshop* 9, 5–36.

Goffman, Erving. 1959. *The Presentation of Self in Everyday Life.* Garden City: Anchor.

——— 1961. *Encounters: Two Studies in the Sociology of Interaction.* Indianapolis: Bobbs-Merrill.

——— 1972. *Relations in Public: Microstudies of the Public Order.* New York: Harper Colophon.

——— 1974. *Frame Analysis: An Essay on the Organization of Experience.* New York: Harper and Row.

——— 1976. *Gender Advertisements.* Washington, D.C.: Society for the Anthropology of Visual Communication.

——— 1979. *Relations in Public.* New York: Harper and Row.

——— 1981. *Forms of Talk.* Philadelphia: University of Pennsylvania Press.

Goodwin, Charles. 1981. *Conversational Organization: Interaction between Speakers and Hearers.* New York: Academic Press.

Hall, Edward T. 1969. *The Hidden Dimension.* New York: Anchor.

——— 1977. *Beyond Culture.* New York: Anchor.

Harcourt, Glenn. 1987. "Andreas Vesalius and the Anatomy of Antique Sculpture." *Representations* 17, 28–61.

Hayles, N. Katherine. 1993. "The Materiality of Informatics." *Configurations* 1:1, 147–170.

Hegel, Georg Wilhelm Friedrich. 1967. *The Phenomenology of Mind.* New York: Harper and Row.

——— [1807] 1977. *Phenomenology of Spirit.* Oxford: Clarendon Press.

Isenberg, Sheldon R., and Owen, Dennis E. 1977. "Bodies Natural and Contrived: The Work of Mary Douglas." *Religious Studies Review* 3:1, 1–17.

Jackson, Michael. 1989. *Paths toward a Clearing: Radical Empiricism and Ethnographic Inquiry.* Bloomington: Indiana University Press.

Johnson, Mark. 1987. *The Body in the Mind: The Bodily Basis of Meaning, Imagination, and Reason.* Chicago: University of Chicago Press.

Joubert, Laurent. 1579. *Erreurs Populaires.* Bordeaux.

Kristeva, Julia. 1982. *Powers of Horror: An Essay on Abjection.* Trans. Leon S. Roudiez. New York: Columbia University Press.

Laboratory of Pathology. 1987. *Autopsy Reports.* "University Hospital."

Labov, William. 1972. *Language in the Inner City: Studies in the Black English Vernacular.* Philadelphia: University of Pennsylvania Press.

Labov, William, and Joshua Waletzky. 1967. "Narrative Analysis." In *Essays on the Verbal and Visual Arts,* ed. June Helm, 12–44. Seattle: University of Washington Press.

Lacan, Jacques. 1977. *Ecrits.* Trans. Alan Sheridan. New York: W. W. Norton.
Laqueur, Thomas. 1990. *Making Sex: Body and Gender from the Greeks to Freud.* Cambridge, Mass.: Harvard University Press.
Leach, Edmund. 1971. *Rethinking Anthropology.* London: Athlone Press.
Leder, Drew. 1990. *The Absent Body.* Chicago: University of Chicago Press.
Locke, John. [1690] 1959. *Two Treatises of Government.* New York: Hafner Publishing Company.
Ludwig, M.D., Jurgen. 1979. *Current Methods of Autopsy Practice.* Philadelphia: W. B. Saunders.
Lyotard, Jean. 1984. *The Postmodern Condition: A Report on Knowledge.* Minneapolis: University of Minnesota Press.
MacCannell, Dean. 1990. "Cannibal Tours." *Society for Visual Anthropology Review,* 14–24.
Marcel, Gabriel. 1960. *The Mystery of Being.* Vol. 1: *Reflection and Mystery.* Chicago: Gateway.
Marcus, George E., and Dick Cushman. 1982. "Ethnographies as Texts." *Annual Review of Anthropology* 11, 25–69.
Merleau-Ponty, Maurice. 1964. *Signs.* Trans. Richard C. McCleary. Chicago: Northwestern University Press.
——— 1971. *The Primacy of Perception.* Ed. James M. Edie. Chicago: Northwestern University Press.
——— 1973. *The Prose of the World.* Ed. Claude Lefort. Trans. John O'Neill. Evanston, Ill.: Northwestern University Press.
Miller, D. A. 1986. "Cage aux Folles: Sensation and Gender in Wilkie Collins' *The Woman in White.*" *Representations* 14, 107–136.
Miller, Jonathan. 1978. *The Body in Question.* New York: Random House.
Mishler, Elliot. 1984. *The Discourse of Medicine: Dialectics of Medical Interviews.* Norwood, N.J.: Ablex.
Morris, Desmond. 1969. *The Naked Ape.* London: Corgi.
——— 1973. *Manwatching.* London: Jonathan Cape.
——— 1985. *Body Watching.* New York: Crown.
Natanson, Maurice. 1962. *Literature, Philosophy, and the Social Sciences: Essays in Existentialism and Phenomenology.* The Hague: Martinus Nijhoff.
——— 1970. *The Journeying Self: A Study in Phenomenology and Social Role.* Reading, Mass.: Addison-Wesley.
Needham, Rodney. 1981. *Circumstantial Deliveries.* Berkeley: University of California Press.
Nicolaisen, Wilhelm F. H. 1991. "The Past as Place: Names, Stories, and the Remembered Self." *Folklore* 102:i, 3–15.

Parsons, Talcott. 1964. *Social Stucture and Personality.* New York: Free Press.

Polanyi, Michael. 1969. *Knowing and Being.* Ed. Marjorie Grene. Chicago: University of Chicago Press.

Richter, D. H. 1986. "Bakhtin in Life and in Art." *Style* 20:3, 411–419.

Ricoeur, Paul. 1971. "The Model of the Text: Meaningful Action Considered as a Text." *Social Research* 38, 529–562.

——— 1980. "Narrative Time." *Critical Inquiry* 7:1, 169–190.

Rimmon-Kenan, Shlomith. 1984. *Narrative Fiction: Contemporary Poetics.* London: Methuen.

Ritchie, Susan. 1988. "Reanimating the Corpus: Points of Slippage in the Ideological Creation of Subjecthood." Paper presented at the American Folklore Society Meetings, Cambridge, Mass.

——— 1993. "A Body of Texts: The Fiction of Humanization in Medical Discourse." In *Bodylore,* ed. Katharine Young, 205–223. Knoxville: University of Tennessee Press with the Publications of the American Folklore Society.

Rose, Dan. 1986. "Transformations of Disciplines through Their Texts: An Edited Transcription of a Talk to the Seminar on the Diversity of Language and the Structure of Power and an Ensuing Discussion at the University of Pennsylvania." *Cultural Anthropology* 1:3, 317–327.

Russo, Mary. 1995. *The Female Grotesque: Risk, Excess, and Modernity.* New York and London: Routledge.

Ryan, Marie-Laure. 1984. "Fiction as a Logical, Ontological, and Illocutionary Issue." *Style* 18:2, 121–139.

——— 1991. *Possible Worlds, Artificial Intelligence, and Narrative Theory.* Bloomington: Indiana University Press.

Sacks, Harvey. 1992. *Lectures on Conversation,* vols. 1–2. Ed. Gail Jefferson, introduction by Emanuel A. Shegloff. Oxford: Blackwell.

Sacks, Oliver. 1985. *The Man Who Mistook His Wife for a Hat.* New York: Summit Books.

Sartre, Jean-Paul. 1964. *Being and Nothingness.* New York: Citadel Press.

Scarry, Elaine. 1987. *The Body in Pain: The Making and Unmaking of the World.* New York and Oxford: Oxford University Press.

Schenkein, James, ed. 1978. *Studies in the Organization of Conversational Interaction.* New York: Academic Press.

Scheper-Hughes, Nancy, and Margaret M. Locke. 1987. "The Mindful Body: A Prolegomenon to Future Work in Medical Anthropology." *Medical Anthropology Quarterly* 1:1, 1–36.

Schilder, Paul. 1950. *The Image and Appearance of the Human Body.* New York: International Universities Press.

Schutz, Alfred. 1973. *On Phenomenology and Social Relations.* Chicago: University of Chicago Press.

Searle, John. 1969. *Speech Acts.* London: Cambridge University Press.

Selzer, Richard. 1976. *Mortal Lessons: Notes on the Art of Surgery.* New York: Simon and Schuster.

Sheets-Johnstone, Maxine. 1990. *The Roots of Thinking.* Philadelphia: Temple University Press.

Shotter, John and Kenneth Gergen, eds. 1989. *Texts of Identity.* London: Sage.

Shuman, Amy. 1986. *Storytelling Rights: The Uses of Oral and Written Texts by Urban Adolescents.* Cambridge: Cambridge University Press.

Sommer, Robert. 1969. *Personal Space.* New Jersey: Prentice-Hall.

Stafford, Barbara Maria. 1991. *Body Criticism: Imaging the Unseen in Enlightenment Art and Medicine.* Cambridge, Mass.: MIT Press.

Stallybrass, Peter, and Allon White. 1986. *The Politics and Poetics of Transgression.* Ithaca, N.Y.: Cornell University Press.

Stewart, Susan. 1984. *On Longing: Narratives of the Miniature, the Gigantic, the Souvenir, the Collection.* Baltimore: Johns Hopkins University Press.

Sudnow, David. 1967. *Passing On.* London: Prentice-Hall.

Tedlock, Dennis. 1978. *Finding the Center: Narrative Poetry of the Zuni Indians.* Lincoln: University of Nebraska Press.

Tillyard, E. M. W. N.d. *The Elizabethan World Picture.* New York: Vintage Books.

Turner, Victor. 1970. *The Forest of Symbols.* Ithaca, N.Y.: Cornell University Press.

——— 1980. "Social Dramas and Stories about Them." *Critical Inquiry* 7:1, 141–168.

Van Gennep, Arnold. 1960. *The Rites of Passage.* Chicago: University of Chicago Press.

Van Maanan, John. 1988. *Tales from the Field: On Writing Ethnography.* Chicago: University of Chicago Press.

West, Candace. 1984. *Routine Complications.* Bloomington: Indiana University Press.

Wollheim, Richard. 1968. *Art and Its Objects.* New York: Harper and Row.

——— 1984. *The Thread of Life.* Cambridge, Mass.: Harvard University Press.

Workman, Mark. 1993. "Paradigms of Indeterminacy." Paper presented at Inquiries in Social Construction, Durham, N.H.

Young, Katharine. 1986. *Taleworlds and Storyrealms: The Phenomenology of Narrative.* Dordrecht: Martinus Nijhoff.

Young, Katharine, ed. 1993. *Bodylore.* Knoxville, Tenn.: University of Tennessee Press with the Publications of the American Folklore Society.

Young, Katharine, and Barbara Babcock, eds. 1994. Special Issue on Bodylore. *Journal of American Folklore* 107: 423.

Zaner, Richard. 1981. *The Context of Self: A Phenomenological Inquiry using Medicine as a Clue.* Athens, Ohio: Ohio University Press.

Zerubavel, Eviatar. 1983. *Patterns of Time in Hospital Life.* Chicago: University of Chicago Press.

INDEX

• • •